MARITIME ARCHAEOLOGY
AND
SHIPWRECKS OFF GOA

MARITIME ARCHAEOLOGY AND SHIPWRECKS OFF GOA

SILA TRIPATI
A. S. GAUR
SUNDARESH

Kaveri Books
New Delhi-110 002

First Published in 2014
ISBN 978-81-7479-140-5

Published by
Kaveri Books
4832/24, Ansari Road
New Delhi - 110 002 (India)
Tel.: 011-2328 8140, 2324 5799
E-mail: kaveribooks@vsnl.com,
kaveribooks@gmail.com
Website: www.kaveribooks.com

Laser Typesetting by : Deep Graphics, Delhi

Printed at : Salasar Imaging Systems, Delhi-110 052

PRINTED IN INDIA

Dedicated to

the mariners

who lost their lives

during sea voyages in

these waters

25 Years of

Maritime Archaeological Research in Goa Waters

understanding the seas

CSIR - National Institute of Oceangraphy, Goa, India

Foreword

Shipwreck exploration was taken up in India for the first time in the waters off Goa in 1988 by the CSIR-National Institute of Oceanography (NIO), a constituent laboratory of the Council of Scientific and Industrial Research (CSIR). Since then, search for shipwrecks along the coast of Goa have been taken up off Sunchi Reef, St. George's Reef, Amee Shoals, Sail Rock and Grande Island. The present book describes the findings from these studies.

The Sunchi Reef shipwreck is datable to the early 17th century and provides insights into the Indo-Portuguese maritime trade of those days. The St. George's Reef shipwreck is 115 years old and is the first Basel Mission Company shipwreck so far explored. The Amee Shoals shipwreck, the first steam engine wreck observed in the waters off Goa, could be a cargo ship of the early 1900s that was owned by a British company. The shipwreck off Sail Rock could be from late 19th or early 20th century. The Grande Island shipwreck could be a century old.

This book on underwater exploration off Goa provides first hand information and analyses of the artefacts that were recovered from the shipwrecks. The book offers an opportunity to study the maritime history of Goa from a new perspective by shedding new light on the underwater cultural heritage of Goa for the benefit of scholars, researchers, and students of archaeology of India in general and Goa in particular.

S. R. Shetye

S.R. Shetye

Acknowledgements

We are extremely grateful to Dr. S. R. Rao, (Shrikaripura Ranganatha Rao) father of Marine Archaeology in India, who pioneered Maritime Archaeological studies in India in 1981 as a project with financial support extended by the Indian National Science Academy (INSA), the Department of Science and Technology (DST), Archaeological Survey of India (ASI). Logistic support was provided by the National Institute of Oceanography (NIO) for undertaking maritime archaeological studies off Goa, Gujarat, and the Tamil Nadu coast. We are obliged to the funding agencies of this great cause. In 1990 the project culminated in the establishment of the Marine Archaeology Centre at NIO, Goa. We are indebted to Dr. B. N. Desai, former Director of NIO, for his continuous support for the establishment of the Marine archaeology Centre at NIO. Subsequently, the Department of Ocean Development (DOD) (Currently known as the Ministry of Earth Sciences) and the Council of Scientific and Industrial Research (CSIR), New Delhi, has funded various projects for which we are thankful to DOD and CSIR. Thanks are also due to the project review committee members namely Shri. R. S. Bisht, former Joint Director General, ASI; Prof. V. H. Sonawane, M. S. University, Baroda; Dr. S. P. Gupta, Chairman of the Indian Archaeological Society, New Delhi; Prof. S. N. Rajaguru, Deccan College, Pune; Dr. K. Rajan, Department of Archaeology, Tamil University, Thanjavur (currently at Pondicherry University) for their encouraging opinions on shipwreck studies off the Goan coast. We are obliged to Dr. E. Desa, Dr. M. D. Zingde, former Directors of NIO and Dr. Satish R. Shetye, the present Director of NIO, Goa for their continuous support for maritime archaeological studies.

This book could not have been written without the published material and ideas of many scholars, discussions with colleagues and associates of India and abroad including our colleagues at NIO, Goa. We are indebted to all our teachers, associates, scientists and friends who showed us the way to approach archaeological problems. Dr Satish R. Shetye, Director, National Institute of Oceanography and Shri K. H. Vora, Scientist-in-Charge, Marine Archaeology Centre has extended all the facilities and constant encouragement which made publication of this work a possibility. Many scholars have provided sincere and involved help for quite some time in this study. It is with pleasure and gratitude that we name a few of them: Ian Godfrey, R. V. Rao, S. Shashikala, G. Rajagopalan, A. R. Gujar, Abhay Mudholkar, P. S. Parameswaran, B. R. Rao, Vijay Khedekar, Cesar Moraes and Girish Prabhu for analyses of ivory and hippo teeth, anatomical study of timber, carbon14 dating of timber, XRD analysis of pottery, thin section and XRF analysis of stone anchors and SEM study of coating material on iron and stone anchors and Patrick Lize, Roxanna M. Brown, N. Karashima and Hanae Sasaki for providing information on shipwrecks on both east and west coasts of Indian waters and identification of Chinese ceramics.

We are beholden to Valerie Fenwick, George F. Bass, Filipe Castro, K.S. Mathew, Robin Piercy, Francisco Alves, Paula Martin, Bruno Werz, Cheryl Ward, Mack McCarthy, Charles D. Beeker, Malcolm Turner, John Riley, Jean Blot, S. K. Mhamai Kamat, Victor Rangel-Ribeiro, Teotonio R. deSouza who have, provided necessary information on shipwrecks, maritime trade, identification of shipwreck finds, anchors and other relevant inputs and suggestions at every time requested. It is indeed our privilege to say thanks to Valerie Fenwick, M. Dileep Kumar, Sridhar Iyer, Antonio Mascarenhas and Franco Fernando for their valuable suggestions on the manuscript. The authors are also thankful to Shriguru Bagi and Sophia Thomas for furnishing information on Chandor excavations and Archaeological Survey of India (ASI) for granting Permission to undertake the exploration in Goa waters.

We have received help from many people who have contributed in one way or another to the book, and are obliged to them. Anthony (Tony) Estrocio and Menin Cabral have provided their boats for exploration and shared their knowledge on shipwrecks off Goa; Florenco Marques informed us about the find of iron anchors off Aguda Bay. The book contains many figures, photographs and half tones. S. N. Bandodker, S. B. Chitari, Arun Mahale, Ravindra Uchil, Shamkant Akerkar, Kamlakar Chitari, Uday Javali, Pramod Pawaskar, Sheikh Ali Karim and Mahesh Mochemadkar have shown tremendous patience and have supplied wonderful prints.

Preface

The worldwide maritime archaeological evidence for ancient water transport is scanty, both in time and in space. It is believed that the earliest voyages were made around 60,000 years ago, and people must have used some kind of floats or rafts for the purpose. The 7516 km long coastline of India enabled the development and progress of seaports and trade centres from ancient times. Rafts, log and basket boats, were used in olden days for exploitation of marine resources in coastal waters. Dating back to prehistoric times and the people of Neolithic, Chalcolithic and Megalithic cultures used the sea. After realising the usefulness of water transport, distinct types of vessels were designed and constructed. Over a period of time they appear on seals, potsherds, coins, amulets, cave and temple walls as well as rock paintings. The history of boatbuilding is as old as the relationship between man and the sea. The mariners of India undertook overseas voyages to different countries after gaining knowledge of the sea's currents, tides, waves, weather and winds. In overseas trade and commerce the ports and trade centres on the east and west coasts of India played a critical role from at least the 3rd millennium BC onwards.

Literary sources, archaeological discoveries, iconographic and epigraphic evidences and accounts of travellers allude to trade and commerce, overseas cultural contacts, and problems faced by mariners and seamen and shipwrecks in India. In order to corroborate the descriptions in literary sources, epigraphical evidences and accounts of travellers, land and underwater exploration are highly necessary. Maritime archaeological exploration and excavation have been carried out along the Indian coast since the inception of the Marine Archaeology Centre at the National Institute of Oceanography (NIO), a constituent laboratory of the Council of Scientific and Industrial Research (CSIR), Goa. In the last three decades significant achievements have been made in maritime archaeological research in India. During this period underwater exploration has been carried out in Dwarka, Bet

Dwarka, Somnath, Porbandar and the adjoining regions, Lakshadweep, Vijaydurg, Poompuhar, Mahabalipuram and Goa. In the early years of marine archaeological research, underwater exploration was carried out in Dwarka, Bet Dwarka and Somnath waters with the objective of locating submerged habitation sites and ports in order to correlate finds with descriptions in the literature. However, the purpose of underwater exploration off Goa was to locate and explore shipwrecks. Goa has been the focal point for this since the establishment of the Maritime Archaeology Centre at NIO because marine and other related records housed in the Goa State Archives, Panaji and India House, Lisbon state that several ships were wrecked in Goa waters due to manmade and natural causes and warfare. In the beginning, the known shipwrecks were explored and the remains were documented. Later on attempts were made to collect information on unknown shipwrecks off Goa coast. In India shipwreck exploration was initiated in 1986-87 off Grande Island and 1989 off Sunchi Reef. However, since 1997 continuous exploration of shipwrecks has been carried out and shipwrecks have been found off Grande Island, Sunchi Reef, St Georges' Reef and the Amee Shoals. Subsequently, shipwrecks have been located off Poompuhar in Tamil Nadu and Minicoy in Lakshadweep waters.

This work is the outcome of onshore and offshore explorations of Goa. The introductory chapter explains the Maritime History of Goa; Trade contacts of Goa with the Arab world; The advent of the Portuguese; Shipbuilding, Shipping and Shipwrecks in Goa waters. The second chapter comprises Coastal Geology and Geomorphology of Goa; Seabed topography; Equipment and Methodology; Location of shipwreck sites; Limitations during exploration. The third chapter is devoted to Shipwreck Exploration in Goan waters: Exploration and excavation of a shipwreck off Sunchi Reef, Exploration of Basel Mission Company shipwreck off St George's Reef; Steam engine Shipwreck Exploration off the Amee Shoals. The fourth chapter details Stone and Iron Anchors from Goan waters. The fifth chapter deals with exploration carried out off Aguada Bay, Baga, Grande Island, Old Goa and Gopakapatana. In addition, a separate comprehensive report of the analysis of ivory and hippopotamus teeth recovered from the Sunchi Reef shipwreck and iron anchors retrieved from Aguada Bay and an anatomical analysis of the wooden stocks of these two iron anchors are included in chapter six. Chapter seven of the book is devoted to the Bibliography and Index.

A few monographs and research articles have been published by scholars and researchers on maritime trade and socio-economic conditions of Goa. However, no attempt has been made to study shipwrecks, ports and ship anchorage sites, and stone and iron anchors of Goa. We hope this book, 'Maritime Archaeology and Shipwrecks off Goa', will provide useful information. Its purpose will be fulfilled if it motivates readers to further study. This is NIO's contribution No. 5423

Sila Tripati, A. S. Gaur and Sundaresh

Contents

Abbreviations

AAE	Arabian Archaeology and Epigraphy
AOB	Archives of Oral Biology
ASI	Archaeological Survey of India
BAIMA	Bulletin of the Australian Institute for Maritime Archaeology
BCD	Buoyancy Compensator Device
CS	Current Science
EI	Epigraphia Indica
IJHS	Indian Journal of History of science
IJNA	International Journal of Nautical Archaeology
IJMH	International Journal of Maritime History
IJMS	Indian Journal of Marine Science
INA	Institute of Nautical Archaeology
JAS	Journal Archaeological Science
JIH	Journal of Indian History
JIOA	Journal of Indian Ocean Archaeology
JMA	Journal of Marine Archaeology
JMA	Journal of Maritime Archaeology
MAC	Marine Archaeology Centre
NIO	National Institute of Oceanography

NRJ	Nautical Research Journal
PIHC	Proceedings of the Indian History Congress
SCUBA	Self Contained Underwater Breathing Apparatus
SDDE	Surface Demand Diving Equipment
SAS	South Asian Studies
SMA	Society for Marine Archaeology
SNR	Society for Nautical Research
UGC	University Grants Commission
VOC	Dutch East India Company
WAMM	Western Australia Maritime Museum

List of Illustrations

List of Figures

List of Tables

1

Maritime History of Goa

Goa is separated from the mainland by two major rivers, namely the Mondovi on the north and the Zuari on the south. It is almost triangular in shape and the Arabian Sea and Western Ghats surround its major portion. Goa has been referred to as 'Nelkinda' by the author of the *Periplus of the Erythraean Sea (Periplus Maris Erythraei)* (60-100 AD) (Schoff, 1974; and Casson, 1989), 'Melinda' by Ptolemy (150 AD), the Greek geographer (McCrindle, 1885), and 'Nekanidon' by Pliny (79 AD). Goa was known to Arab mariners and traders as 'Kuwa' or 'Kuwe' (Lee, 1829). Al-Eidrissi, Al-Masudi, Rashid-ul-din and Ibn Batuta have referred to 'Sindabur' or 'Sandabur' and scholars have believed Chandor or Chandrapura could be Sindabur (Moraes, 1990; and Fonseca, 1986). Among all the ports and harbours of the west coast of India, Goa occupies a strategic position whilst relatively good harbours have made it a centre of maritime trade for much of its history. Goa was one of the most important entrepots on the west coast of India. Its maritime contacts with other ports developed in the early centuries of the Christian era if not earlier. Goa was associated with inland neighbouring dynasties through the Western Ghats. The maritime history of Goa dates back to the Mauryan period. Chandrapura (modern Chandor) (Fig. 1) in Salcete taluka is located 14 km east of Margao, on the Paroda or Kushavati River, a tidal tributary of the Zuari, approximately 30 km from the sea is thought to be the oldest port of Goa and one of the ancient towns of the Konkan coast. The River Paroda was navigable for large boats until the end of the 19th century but now it is usable only for local craft.

The Siroda copper plates are the earliest to mention Chandrapura (Chandor). These copper plates have been dated to the 4th century AD on palaeographic grounds. They comprise a royal charter issued by King Devaraja of the Bhoja family announcing a

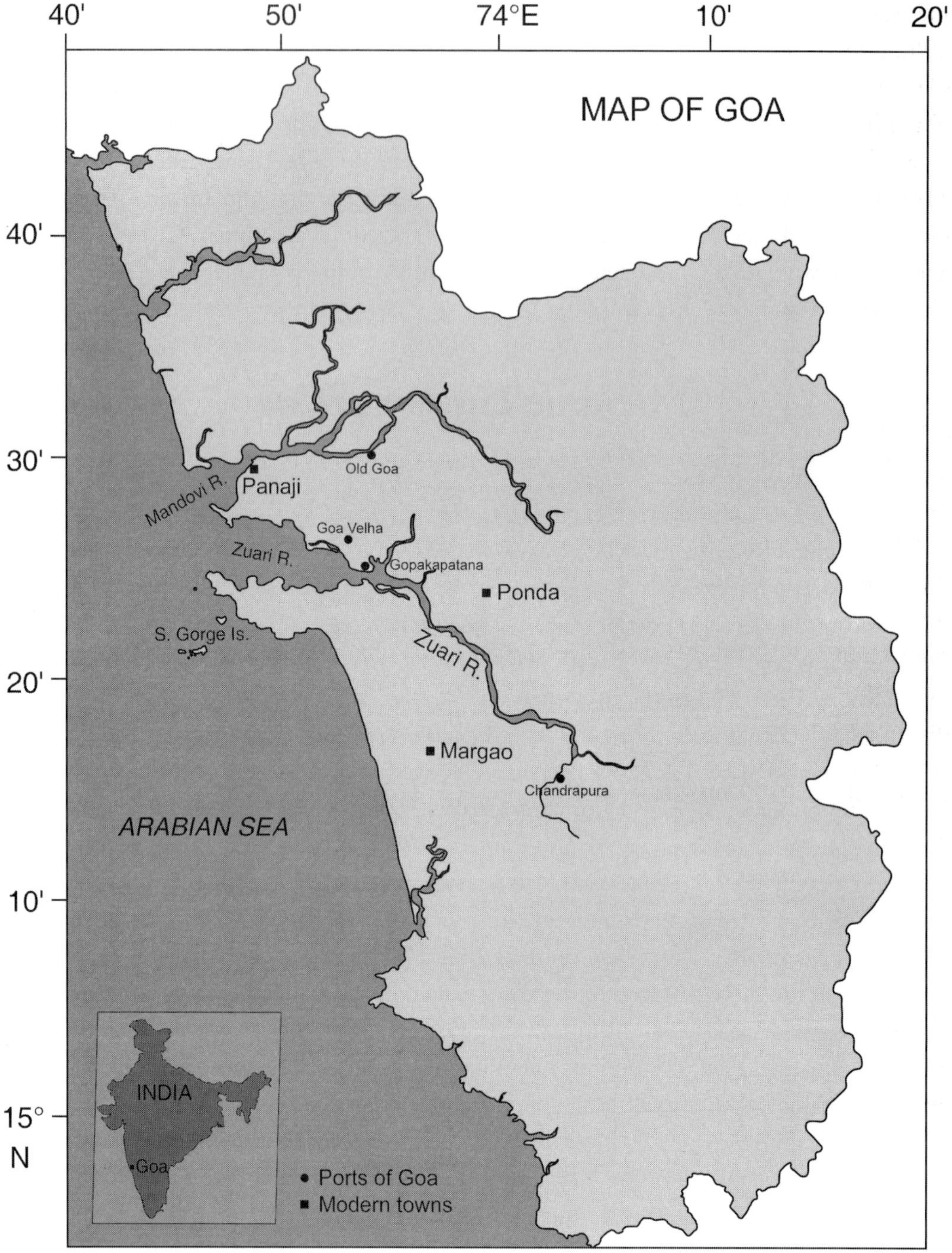

Fig. 1: Location of Chandrapura (Chandor) and other ports of Goa

meritorious benefaction from his capital Chandrapura (Krishnamacharlu, 1942; and Rao, 1952). The copper plate of King Chandravarman of the 5th century AD also refers to Chandrapura. The late 10th century AD Marcella copper-plate inscriptions mention that Shashthadeva II, the Kadamba king established his capital at Chandrapura which is claimed to excel even the metropolis of the gods (Moraes, 1990). The Kharepatana plates of the 11th century AD indicate that ships once docked at Chandrapura. In the 1920s Stuart Gomes declared that Chandor must be the site of ancient Chandrapura. Chandrapura was a fortified town with a mud defensive wall and a moat running parallel to the fortification. Chandrapura was the capital and maritime centre during the reign of the Kadambas (950-1300 AD).

Beside Chandrapura the other flourishing ports were Gopakapatana (Gopakapattana) and Vallipatana. Vallipatana has been alternatively identified as Bali in Kepem (Quepem) and Velim in Sashti (Salcete), Goa, Valavli in Sawantwadi and Kharepatana near Vijaydurg in Maharashtra. Ships loaded with cotton goods, diamonds, precious stones, spices and perfumes sailed to African and Arabian countries from these ports (Shirodkar, 1990) and merchants of Goa dealt in fine muslins, jewels, pepper, spices, camphor, betel nuts and perfumes. The historical evidence shows that Chandrapura was ruled by the Satavahanas, Bhojas and Chalukyas of Badami (580-750 AD). Maritime trade and commerce flourished under the Satavahanas. In the 2nd century BC the Satavahanas overran the Konkan with the Bhojas probably accepting their suzerainty. After the Bhojas, the Satavahana rulers seem to have continued seaborne trade from the ports of Konkan. The *Periplus of the Erythraean Sea* (60-100 AD) mentioned that the south Konkan coast was infested with pirates and hence foreign ships were escorted by Goan mariners to the port of Bharakuccha during the Satavahana dynasty (Schoff, 1974). Over the next few centuries, Goa was ruled successively by the south Konkan Shilaharas, Rashtrakutas, Kadambas and Chalukyas of Kalyani. During the 8th century, it was ruled by the Shilaharas who were in a feudal capacity under the Rashtrakutas (Desai, 1965). Gopakapatana served as the capital and port of the Shilaharas and Vallipatana port was also developed during the Shilaharas period. The Shilahara (750-1010 AD) and the Kadamba (950-1300 AD) dynasties were great maritime powers and maintained trade contacts with east African and west Asian countries (Gune, 1990). The earliest reference to navigation in the River Paroda or Kushavati appears on the Panjim copper plates of Jayakeshi-I (1052). This inscription further vividly describes the city of Chandrapura. It is stated that it comprised white plastered houses, alleys, stables, flower gardens, bazaars, parlours, and quarters. The Kadamba king Jayakeshi-I ruled Chandrapura until the middle of the 11th century and subsequently shifted the capital to Gopakapatana (Moraes, 1990).

Natural processes such as sedimentation, erosion and other geological phenomena caused the decline of Chandrapura on the bank of the River Paroda or Kushavati and this forced the Kadambas to shift their port and capital to Gopakapatana. The Gopaka copper plate mentions that King Viravarmadeva, elder brother of Jayakeshi-I, shifted his capital from Chandrapura to Gopakapatana on the bank of the River Zuari (Gune, 1979) around 1049 probably because of silting of the River Kushavati. Before the capital moved from Chandrapura to Gopakapatana, Gopakapatana already served as a port and trade centre. The finding of Roman coins (309-337 AD); coins with the St Thomas Cross inscribed in 'Pahlavi' (7th century AD) used by Persian traders during the Sassanid dynasty; and sherds of Chinese ceramics indicate that Gopakapatana was an early maritime trade centre (Costa, 1997). Costa believed that Roman ships called at the ports of Goa from the Satavahana period onwards. Sanaphulla, the founder of the southern Konkan Shilahara dynasty, built up Gopakapatana city around the hillock of Pilar after seeing its port frequented by foreign ships.

During the rule of King Jayakeshi-I (Gune, 1979) the Kadambas had reached their zenith of power. Some of the inscriptions of the Kadamba dynasty and the contemporary literary sources refer to sea voyages and naval activities of the Kadamba kings. King Jayakeshi-I organised a strong naval fleet at *Velakula* in Gopakapatana and appointed Saddam (Chhadam), the *Nauvittaka* owner of the Arab merchant ships, as '*Maha Pradhan*' in his navy due to his experience in sea trade and navigation, setting aside other ministers. During this time Arab traders had settled in Goa. The Panjim copper plate of King Jayakeshi-I mentioned that Saddam had established a *mijigithi* or charity house to look after orphans and the destitute which was maintained from the profits of Gopakapatana port (Moraes, 1990). The Panjanakhani stone inscription (1054) mentions King Jayakeshi-I known as *Padaval-endra* which signifies *Paschima-Samudradhishvar* (Lord of the Western Ocean) (Gune, 1979). During the reign of Jayakeshi-I Gopakapatana attained prosperity as a flourishing port and commercial centre. It was one of the most prominent emporia on the west coast of India. During the Kadamba rule Goa possessed a wealthy trading and industrial class which often dealt in gold, silver, spices, fruits and pepper. The trade of the Kadamba dynasty was mainly in the hands of indigenous, itinerant and foreign dealers. King Jayakeshi-I had trade contacts with Simhala (Sri Lanka), Kedah (Sumatra), Zangavar (Zanzibar), Callah (Kuwait), Pandu (Tanjore), Gaudda (Bengal), Gurjara (Gujarat), Srytam (Socotra) as well as Cholas and Pandyas and ships plied from Gopakapatana port to these places. Pissurlencar (1938) states that one of the copper plates kept in Portugal mentions the construction of a mosque at Gopakapatana and for its maintenance King Jayakeshi-I levied taxes on the merchant ships of Simhala, Zangavar, Gaudda and Gurjara which called at the port. This clearly indicates that Gopakapatana was the main port and emporium on the west coast of India during the reign of King Jayakeshi-I and that Chandrapura served as a minor port. Gopakapatana was connected with Ela (Old Goa) by the highway called 'Vidhi' or 'Rajbidi' (Shirodkar, 1990) and 'Rajvithi' or 'Rajpath' still exists (Kamat, 1990). The road was used by traders to convey their cargo to Ela port. The Ganadevi (near Surat in Gujarat) stone inscription of Jayakeshi mentions that he had constructed a market place (Madapika) near Ganadevi port, which shows the interest of Jayakeshi in promoting trade and commerce with other countries.

The inscription of Narendra describes King Jayakeshi-II (1104-1148) as the Konkan Chakravarti (Ruler of the Konkan) who had built a bridge with a line of ships reaching as far as Lanka and had conquered Lanka and Kavadidvipa. Kavadidvipa has been generally accepted as Kavaratti Island in Lakshadweep (Barnett, 1982). The inscription reads as follows: '*Chattayyadevam devaraja prakatita vibhavam Svikritaschayya saurayyam Kavadi dvipamum adiage palavum dvipamgalam kondu*', The English translation is: '*He was Chattayadeva, the exalted Lord of the Western Ocean, who displayed the majesty of the King of the Gods and possessed miraculous heroism. As he took Kavadi-dvipa and many other regions under his control and built a bridge with lines of ships reaching as far as Lanka (and) claimed tribute among grim barbarians, exceedingly exalted was the dominion of the Kadamba sovereign, which many called a religious estate for the establishment of the worship*'. This shows that rulers of the Kadamba dynasty maintained an efficient and strong navy. In addition to the inscriptions, the literary sources describe the construction of ships by the Kadamba kings for maritime trade and naval warfare.

Maritime trade and commercial prosperity in Goa during Kadamba rule can be evaluated from archaeological evidence, inscriptions and finds of gold coins at port sites. Trade was regulated by the guilds, the community dealing with trade was the '*vanias*'(*banias*), and trading settlements were known as '*Hanjamannagar*'. Moreover, the maritime history of Goa is also reflected from the *hero stones* exhibited in the Archaeological Museum (ASI), Old Goa. Of several *hero stones*, ship depictions are found on four. These *hero stones* illustrate naval actions, and probably these attacks took place during the 12th century because Goa Kadambas launched a series of attacks on North Konkan during this period (Rajagopalan, 1987). Two types of ships are represented and they resemble dugouts and canoes. The stem and stern of two vessels are curved like dugouts and they resemble the double-ended canoe-shaped fishing boats working along the Konkan coast. Other boats, however, show long horizontal thick strips placed over the plank seams, evidently a wadding of coir which compresses them and prevents leakage. These two vessels resemble warships called '*gurab*', which were used by the Marathas along the west coast of India. They appear to be indigenous because of their shape and size, and they have been used along the coast for a long period. The vessels depicted on the *hero stones* are warships which were required to move fast and manoeuvre easily. They move fast with the assistance of rowers and in a favourable wind sails could be unfurled. Further, it shows that during the Kadamba period a similar type of craft was used for maritime and naval warfare (Sila Tripati, 2006a). In the recent past a number of *hero stones* with boat motifs have been discovered in the Sattari taluka which provide additional information on the maritime history of Goa (Kamat, 2008). These boats resemble local craft such as canoes which ply coastal and inland waters, and all these boats are represented with a mother goddess. From the design and pattern of the sculptures the *hero stones* can be dated to the medieval period (between 10th and 13th centuries) or later (Gaur and Kerkar, 2007).

Excavation

Chandor has been the capital city, main centre of trade and commerce, as well as a port town for centuries, from the beginning of the early centuries of the Christian Era to the medieval period. Inscriptions and contemporary literature shed light on the socio-economic condition of the people in the region. Several excavations have been carried out at Chandor and in adjoining areas in order to locate habitation sites and any maritime structures. Excavations at the temple site of Chandrapura brought to light evidence assignable to the Satavahana period. It includes Satavahana bricks, pottery, and copper coins datable to the 2nd century BC and a Hindu god and goddess, *Jaina Thirthankars* and coins of the Kadamba period (Heras, 1932). Subsequent excavation by the Archaeological Survey of India (ASI) between 1974 and 1975 yielded bricks and pottery. On the basis of pottery the site has been assigned to the 3rd-4th centuries AD. Subsequently, the Archaeological Survey of India and the Society for South Asian Studies jointly carried out excavations between 1999 and 2002 at the temple site of Chandor in order to understand the nature of occupation during the early period (Kennet and Rao, 2001). Excavation revealed five phases of construction on the site. A trial pit was excavated on the eastern side of the temple to investigate building activities outside it. The upper deposits have three levels of floors consisting of laterite and rammed clay topped by packed tile flooring. The surfaces are very hard and perhaps functioned as a main street outside the enclosure wall. Beneath these deposits a fireplace with charcoal was found, indicating that the area might have been used for religious activities. The discovery of a fireplace near the temple has more significance. Further excavation below the fireplace brought to light a floor laid with tiles (IAR, 2001-02).

Excavations were also carried out near Santiago Chapel, Bamanpura (Bamanura), Chapel of Holy Cross, Rampart East and North, and the temple site of Chandor. In the course of the work bricks, pottery and tiles datable to the 7th century were recovered from the Santiago Chapel site. Trial digging behind Sara Fernandes house yielded remains of the Kadamba period datable to the 9th century as well as later. The significant find is an embankment 6 m long made of *trap stones* found 30 m away from the present river course. Frequent floods had destroyed the major area of Chandor and the River Kushavati became silted up - hence the capital was shifted from Chandor to Gopakapatana. The regular occurrence of floods probably necessitated construction of the embankment, which also facilitated the loading and unloading of cargo. Another trial-pit in the adjoining area of Sara Fernandes house yielded a number of crucibles and copper slag, evidence for copper smithing in the Kadamba period. It is interesting to note that the locality of Chandor is known as 'Kamarkund', meaning smith's furnace. Small digs in the Bamanpura area brought to light pre-Portuguese and Portuguese pottery and trial digs around the eastern side of the rampart and riverbed produced pottery (IAR, 2001-02).

Furthermore, excavations have revealed that the fortification wall was built in phases. In the first phase of construction, clay was used, whereas bricks and brickbats were used in the second phase. Subsequently, river pebbles were used on the outer side of the fortification wall. This shows that the fortification wall was constructed with both brick and stone facings and its height was 8 m and width 8 to 10 m, but information has been lost due to vandalism and levelling for cultivation. An area around the Santiago Chapel has been identified as the entrance to the fort. In addition to structural remains, types of pottery, namely, dull red ware, burnished red ware, black ware, grey ware and glazed ware probably of west Asian origin, were encountered during the excavation. A wide variety of porcelain, sherds of glazed ware and different coloured glass of Chinese, Islamic and Portuguese origin were recovered. Other finds were glass bangles, semiprecious stone ornaments, stone sculptures of recumbent *naga images*, a seated goddess and gold items such as rings and Portuguese coins. Most of the pottery and other finds from Chandor belong to the early historical and medieval periods (Personal communication: Shriguru Bagi).

Cosme (Costa, 1997) carried out surveys around Gopakapatana and in the Pilar region to locate any remains of the port and antiquities. In the course of investigations Roman amphorae, a bust of a Ganesh idol, pottery, stoneware and glasses were collected. The removal of silt and debris from the tank at Pilar brought to light two Roman coins, Kadamba coins, Chinese ceramics, *hero stones* and other artefacts. These Roman coins belong to Constantine the Great (309-337 AD). The discovery of the Roman coins indicates that Roman ships called at ports of Goa during the Satavahana period. Moreover, Chinese ceramics and Arab pottery have been found in the adjoining area of Gopakapatana. These finds emphasize the role of Gopakapatana as an ancient port. All the artefacts from Gopakapatana and the adjoining area have been put on display in Pilar Museum (Costa, 2002). Onshore exploration and trial scraping were carried out to trace the earliest evidence at Gopakapatana (Fig. 2). In order to ascertain the continuation of Rajbidi or Rajvidhi three trial pits were dug at the highest waterline point near the road to Donda village. Each pit was deeper than one metre and brought to light laterite blocks, lime mortar, humus and gravels with reddish sand (Gaur, 1992). Apart from this no other archaeological remains were encountered. The finds may date part of the Rajbidi to the Kadamba and Shilahara periods. Onshore exploration brought to light laterite walls between Gopakapatana and Agasaim which run both perpendicular and parallel to the River Zuari. In the construction of the walls well-dressed laterite blocks of different sizes were used. A wall 1.5 km long is exposed during low tide at Gopakapatana. In some places the wall is either missing or buried in the sediment. Several bastions at regular intervals along one of the walls and six stone steps leading to the shore were noted and these steps are integral with the main wall. All the stone structures lie very close to the royal road (Rajbidi) of the Kadambas. The exploration at Gopakapatana so far has yielded no finds to date these structures (Sila Tripati et al; 2008). Similar structures have been noted all along the coastal area of Goa and therefore the question arises whether the structures of Gopakapatana belong to a port as well as to the Kadamba period. Large-scale survey and excavation would help elucidate the stone structures and their purpose.

Trade contacts of Goa with the Arab world

The merchants of Arabia, Persia and Hormuz brought horses, perfumes and Chinese ceramics to Goa and in turn they took rice, betel, spices, textiles, sugar and fine muslin. On the west coast of India, Goa was the main centre for importing horses and the starting point for Mecca bound pilgrims. An incident is related that Brahmin merchants of the Hassan district were involved in importing elephants, horses and pearls in ships and selling them to the kings. It is not known whether they had direct contact with Arab merchants or imported them from Arab countries (Moraes, 1990). Horses were in great demand and essential for maintaining the military strength of the empires of Bijapur, Golkonda and Vijayanagar (Kamat, 2000). There were frequent battles between these dynasties for supremacy; hence Arab horses were highly necessary. Arabs knew Goa as Sandabur or Sindabur, but some scholars have identified Sindabur as Shadashivagad (Nainar, 1942), whereas others have identified it as Chandrapura and Gopakapatana (Moraes, 1990; and Costa, 2002), added to which the Arabic word 'Sindabur' seems to have been a corruption of 'Chandrapura'

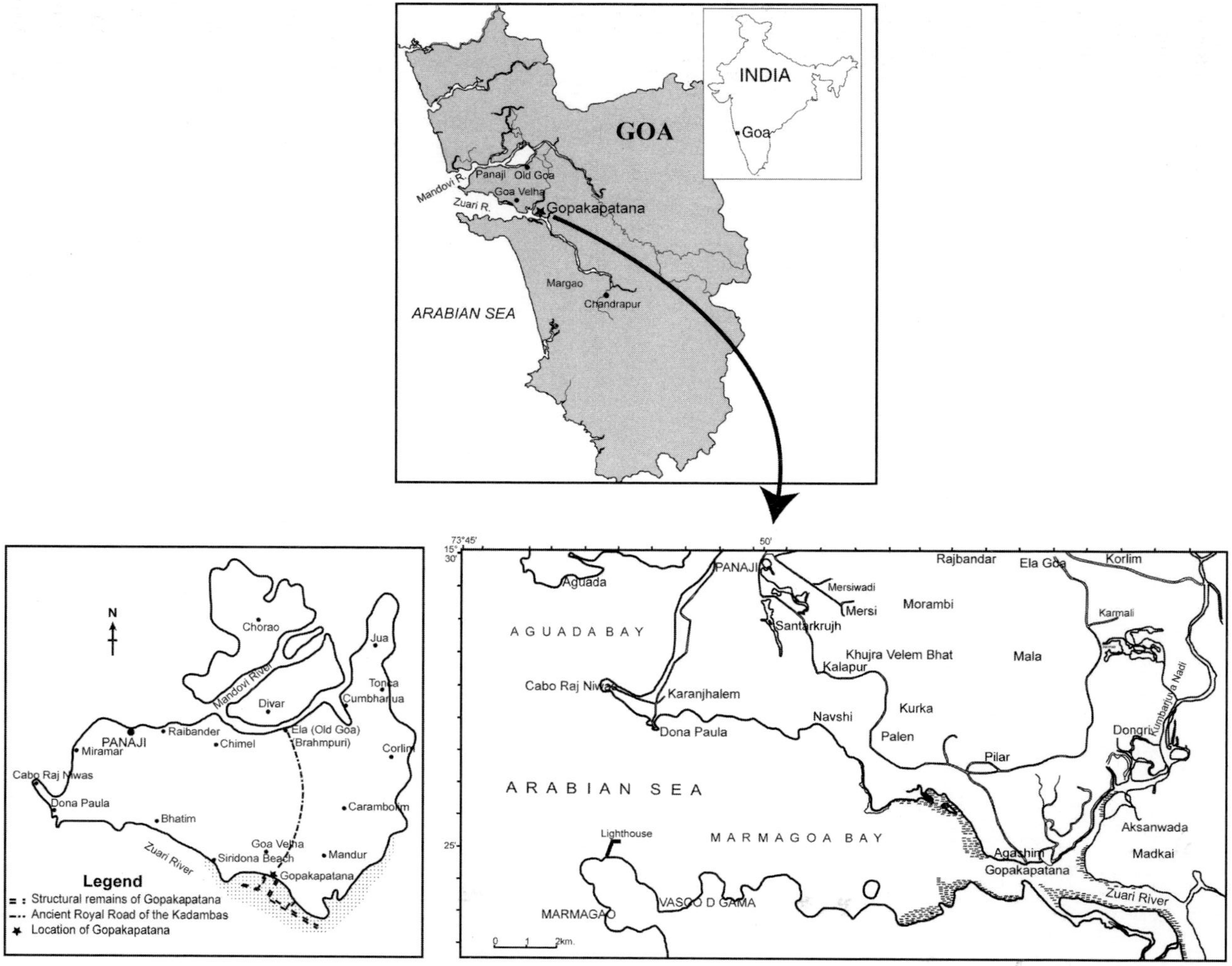

Fig. 2: Location of Gopakapatana and surrounding region and the ancient road of the Kadambas from Gopakapatana to Ela (Old Goa)

(Heras, 1932). Gopakapatana was the entry point for Arab traders *en route* for the Bijapur and Vijayanagar kingdoms. The degree of trade and commerce in Goa during the Kadamba period is evidenced by the discovery of gold coins at Chandrapura, Gopakapatana and other places.

Goa fell to the Muslims for the first time in 1312. Mallik Kafur, the General of Alaudin Khiliji, overran the whole of Goa and looted and destroyed the capital, Gopakapatana. This caused the Kadambas to re-establish themselves at the more secure Chandrapura until in turn it was destroyed by Muhammad bin Tughluq in 1327 (Moraes, 1990; Gune, 1979; and Gibb, 1929). Even Ibn Batuta mentioned the preparations of Jamal Oddin who set out to lead an expedition against Sindabur, and the resistance made by people of Sindabur. Further, refers to the Hindu city of Goa on the bank of the River Zuari and the Mohammedan city of Ela on the bank of the River Mandovi (Fonseca, 1986). Chandor did not return to its original state after its destruction by Muhammad bin Tughluq and remained a small village. Muslim invaders were forced to evacuate in 1370 by Harihara I of the Vijayanagar Empire. Goa witnessed a cultural revival and prosperity during Vijayanagar rule in 1369. Rulers of the Hindu stronghold, Vijayanagar, dominated Goa for nearly 100 years. One of the striking features of Vijayanagar rule was the development of ports and guilds. During this period Gopakapatana became an important port for ships carrying Arab horses for the Vijayanagar cavalry. In the latter part of Vijayanagar rule, Gopakapatana lost its commercial importance due to silting of the River Zuari. It became too shallow to provide shelter for big ships. Vijayanagar rulers also developed Gove Bandar, probably known as Raibander. It served as an emporium for the Vijayanagar Empire and a gateway of lucrative trade with Arabian and western countries between 1367 and 1472 (Desai, 1970). Consequently, Vijayanagar rulers signed a treaty with the Portuguese to provide facilities for the importation of Arab horses via ports of Goa.

In 1469 Goa was conquered by the Bahmani sultans of Gulbarga. During Bahmani rule the capital of Goa was once again shifted from the banks of the River Zuari to the banks of the Mandovi at Ela (the present Old Goa) in 1472. Yusuf Adil Shah of Bijapur conquered Goa 18 years later and Ela became his capital and also served as an important port and trade centre in the Indian Ocean. Adil Shah requested Afonso de Albuquerque to permit him to purchase horses at the port of Goa because Arab horses served in the armies of the central and southern kingdoms. The palace of Adil Shah (the old secretariat building of Panaji) was taken over by the Portuguese viceroys as their official residence. During Portuguese rule Old Goa became the main commercial city. The hinterland was connected by the land and riverine routes from various ports. Merchants from Arabia, Mozambique, Java, Sri Lanka, Maldives and China visited for trade and commerce. Despite these political changes, Old Goa continued to be a major port and its population remained overwhelmingly peasant and Hindu.

The advent of the Portuguese

Before the arrival of the Portuguese in India both occidental and oriental trade were monopolised by the Arabs. They did this by sailing goods from India up the Persian Gulf to Hormuz and transporting them, thence overland to the continent. During the immediate pre-Portuguese period Goa was under the Deccan Sultanate of Bijapur. The Sultanate had a strong control over trade and commerce and kept a strict vigil on the movement of people. In order to control movement a captain was appointed by the king and nobody was allowed to enter the city of Goa without a valid pass (Barbosa, 1967). During this period Old Goa was visited by traders of various nationalities. The book of Tomē Pires (1944 and 1967; Cortesão, 1990 and 2005) details trade and commerce in Goa and adjoining regions at the time of the arrival of the Portuguese. The port of Hormuz at the mouth of the Persian Gulf traded with Aden, Cambay, the kingdom of Deccan (Bahamins), the kingdom of Vijayanagar and Malabar. Horses were imported from Arabia, Hormuz (Ormuz) and Persia to be sold for high prices there and in Deccan and Vijayanagar. Good horses cost 700 *xerafins* (Cortesão, 2005). Overland caravans bought the fabled riches of southern India by exchanging horses and muskets. Furthermore, Pires remarked that people in Goa were most skilled in shipbuilding; new docks built at Ela on the banks of the River Mandovi were capable of providing shelter for seagoing ships. The shipbuilding industry of Goa flourished because of the availability of good quality timber and craftsmen. Ships built there were highly valued and favoured everywhere. The seamen and merchants of Goa were known to be industrious and capable. During this period pepper, ginger, cloves, cinnamon and other spices were imported from Malabar and these products were exported to Arabia, Hormuz and Aden along with other items such as betel and rice. Goa produced the best variety of betel which was praised in Aden, Hormuz and Cambay. The port of Goa and the surrounding regions yielded 400,000 *pardaos* a year from the customs duties on everything that came into the port including other revenue from local produce (Gune, 2002).

Ships laden with coarse and fine cotton cloths, rice, sugar, spices and gems of inferior variety travelled to Socotra every year. Ships also sailed to India via the Arabian coast with following southwest wind as far as Rasul-Hadd and then followed the Persian and Makrani coast to Sindh and Cambay, thus avoiding the full force of the monsoon. It was a place where there was much trade in a variety of commodities. Ships from Aden, Cambay, Mekkah, Hormuz and other countries called at the port. Ela town was bigger with wide streets surrounded by a fortification wall and bastions (Fonseca, 1986). During this period merchants from Arabia, Mozambique, Java, Sri Lanka, Maldives and China visited Goa for trade and commerce. Old Goa was connected by land and riverine routes. On the land route traders went up to the Bijapur and Vijayanagar kingdoms by crossing the ghats. This caravan route was used for the transport of horses and materials. The ghat route enabled Goa to serve the hinterland as an efficient outlet (Pinto, 1994).

The history of Goa and the west coast of India witnessed a complete change after the arrival of Vasco da Gama at Kapukad (Kappat), a few kilometres north of Calicut (Kozhikode) in Kerala on 17th May 1498 on the *Sãn Gabriel,* in company with the *São Rafael* and the *Bérrio* and a store ship. Portuguese ships were the first men-of- war in the Indian Ocean on which guns were mounted. The *Sãn Gabriel* had 20 guns. The guns and other armament on Portuguese ships had a strong effect on Indians. The policy of the Portuguese was to control the whole trade of India and to intimidate to the Indians by showing naval strength. On his second voyage to India Vasco da Gama brought twenty armed ships with the intention of raising commercial war against the traders and rulers of India in order to eliminate the Arab, Gujarati, Karimi, Turkish, Italian and Catalan merchants involved in the spice trade with India. Maritime trade in Goa entered a new phase in 1510 when it was taken over by Afonso de Albuquerque (1453-1515 AD), who conquered Adil Shah of Bijapur. He was able to do this because the Portuguese were invited and supported by Timoja. Timoja was determined to get rid of the Mohammedan yoke (Esteves, 1966). Albuquerque planned to make Goa a suitable port and develop it according to his own ideas and design. He fortified it and made it the seat of the Portuguese trade centre. Albuquerque was the first foreigner who occupied Indian territory without coming via the Hindukush-Khyber route. Subsequently, he captured strategic points, such as Hormuz which controlled the Persian Gulf and Diu, the entrance to Gujarat. Goa dominated the Deccan region, Cochin being the main centre of the pepper trade, Colombo was the entrepot for cinnamon and lastly, Malacca was the gateway to Southeast Asian countries. Thus Arab sources were cut off. Portugal took the reins of trade and empire until the Dutch, English and French made inroads. Afterwards the world changed and Goa receded into history as an entrepot par excellence. The whole region which Albuquerque brought under his control enabled him to effectively dominate the monsoon trade routes in the Indian Ocean and Persian Gulf and enabled him to pursue his own policy. The foundation on the eastern seas continued for 450 years. Economic and military powers rested on the strategic ports through which trade passed. The Portuguese were intent on setting up a string of coastal stations to the Far East in order to control the lucrative spice trade. Goa was the ideal base for the seafaring nation because of its natural harbours and navigable rivers. The Portuguese needed a secure base from where they could patrol the whole western coast of India and Goa was located in the central region of the Arabian Sea coast, midway between the key economic areas of Malabar and Gujarat. Portuguese commercial and political dominance was due largely to their naval superiority. They were successful against Indian rulers but not a match for the Dutch, British and French against whom they fought a number of naval battles in the Indian Ocean.

Amongst all the Asian and the European rulers of India, the Portuguese ruled over Goa for the longest period - nearly four and a half centuries. It is a fact that they administered only a tiny portion of the Indian subcontinent, but it is they who were the first to introduce colonial rule in India, a rule which also lasted until the middle of the twentieth century. During their rule, Old Goa was their headquarters and a prosperous commercial city, both internal and external trade was receiving an exceptional impetus. In order to improve administration and to facilitate trade, the Portuguese created several administrative and religious establishments in Goa. They relied on sea power and trade, and ceased internal expansion as it would not yield any additional economic profit. The desire of the Portuguese was to acquire products from the African coast, sell in the East, crush Arab power and trade as much as possible. They entered into agreements with local merchants and the rulers to obtain articles which were sent to Portugal. The King of Portugal reserved to himself the monopoly of spices which yielded high profits (Fonseca, 1986). The commercial motives were that exotic and aromatic goods, spices, pepper, cardamom, indigo, saffron, pearl, ivory, silk, muslin, alum and carpets, fetched good revenue. Coffee, tea, liquors, gunpowder, finished ivory artefacts, sandalwood, sugar, textiles were exported and slaves, tobacco, ivory, Chinese porcelains and gold were imported. Ships were sent from Goa to Diu, Bahrain, Basra, Muscat and other places loaded with spices, lead, coconuts, ivory and drugs. *Rua Direita* or *dos Leiloes* was the chief trade centre of Old Goa (Mathew, 1990). There were several streets in which craftsmen had their shops and separate streets were designated for certain types of commodities. Items brought from Bengal, Gujarat, Sind and other places were available in these streets. Merchants of varied religions and from different parts of India and abroad were to be seen in Old Goa. In addition to merchants and traders, a number of brokers and bankers were also found in the city (Linschoten, 1903).

After the shift of the headquarters of Portuguese India from Cochin to Goa in 1530, most of the trade and commerce was carried out from there. Goa had direct maritime trade contacts with other countries, for instance, Mozambique, Mombasa and Basra apart from Portugal (Bragança Pereira, 1937). It became the capital of the Portuguese empire in the East and was granted the same civic privileges as Lisbon. It reached its greatest splendour between 1575 and 1600, the age of Golden Goa, but when the Dutch began to control trade in the Indian Ocean their power declined. During the 16th century a number of administrative and religious establishments were constructed in Old Goa. The city of Goa remained the hub of Portuguese commerce in India until the end of the 17th century. Subsequently they shifted their capital to Panjim (Panaji) in 1754. The fall of the Vijayanagar Empire in 1565 caused the decline of lucrative trade between Goa and the Hindu kingdom. Once the Portuguese became supreme over Goa and other regions all foreign and native ships were allowed to ply only under their *Cartaz* (passport system). In this manner they introduced a controlled navigation system. It effectively helped to establish their maritime supremacy on trade in three ways: (a) Indian vessels plying the coast were forced to pay tax and obtain a licence (b) some of the trade-routes were under the monopoly of the king (c) certain merchandise was earmarked for export only in Portuguese ships. The Portuguese even claimed the right to control and tax all other trade in the Indian Ocean. The *Cartaz* made a great impact on Asian trade and was imposed with the help of their strong navy. The Portuguese stationed armed with arms and ammunitions vessels to check the entry of enemies from south and north. On the northern coastline naval vessels were stationed at Bassein, for the southern coast at Cochin, whilst a naval armada was established in Goa as the headquarters. The *galeotas* and larger vessels called *gales* comprised the naval fleets stationed in Goa under the command of the Capitão-mores (Chief Captain). After capturing Goa, Afonso de Albuquerque found it was a wealthy place with vital trade, industry and military importance.

And at Goa port Albuquerque found a thriving dockyard in the captured city which he named as '*Ribeira*' or *Ribeira-das-Gale's*. The dockyard of Old Goa had about 40 ships, 26 *brigs* and a large number of *fustas*. There was a large amount of artillery comprising 40 guns, 55 ordnances (falcons), 200 muskets and a large quantity of gunpowder and naphtha oil. The port of Goa was well developed and adequate infrastructure facilities were available. The city of Old Goa was flourished with commercial activities; it had the Viceroy's palace, a bazaar and churches (Fig. 3). Old Goa port had two anchorages, the Quay of the Viceroys and the Quay of Catherine. The former was located near the Arch of the Viceroys on the banks of the River Mandovi. Ships of different countries fully loaded arrived and after unloading, the cargo was taken to '*Bangacal*', the store building. There were other large buildings close to the quay where cargo was stored and weighed. It was also the landing place for passengers. *Alfandega* (Customs House) was located on the south side of St. Catherine's Chapel. These were built mainly for the use of the viceroy and other high dignitaries. The Quay of Catherine was also called *fish bazaar* and lay in front of St. Catherine's Chapel. It was located north of the Royal Hospital, extended as far as the Arsenal and was the main place for landing and embarkation. '*Ribeira-da-Gales*' was the mooring place for the galleys. Access to this area was only on business. Ships sailing to

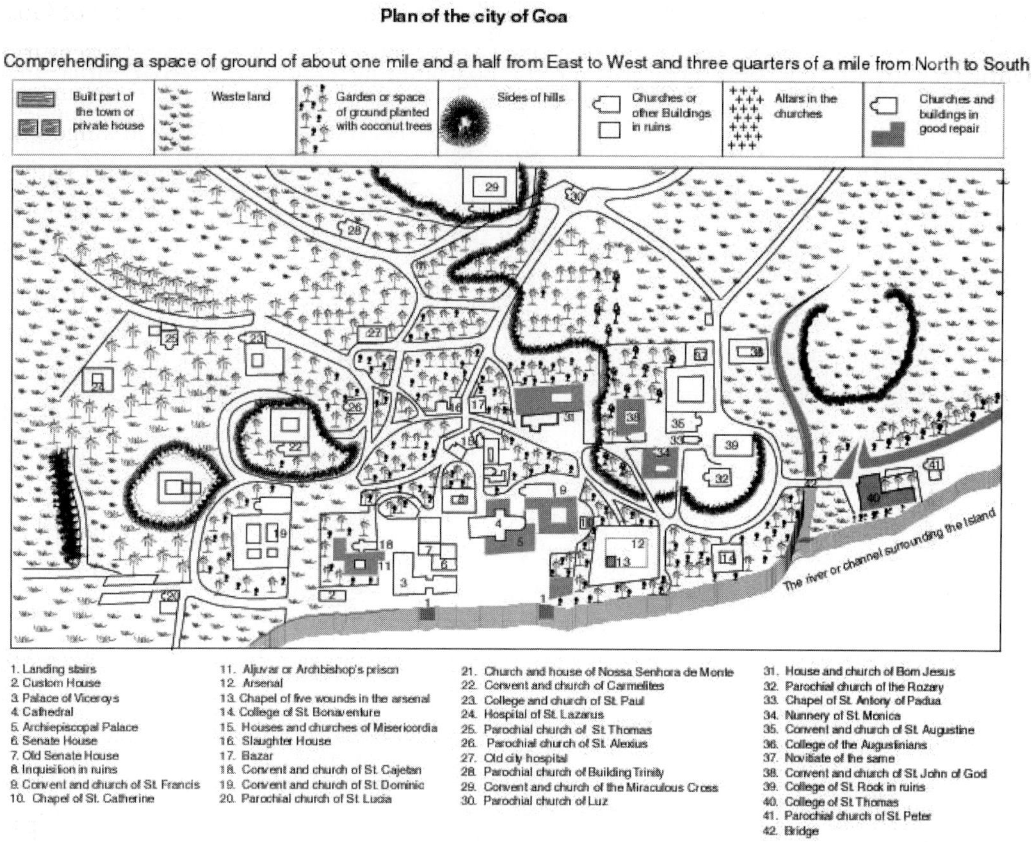

Fig. 3: Plan of Old Goa city and buildings in the surrounding area

Portugal were loaded here with merchandise (Nambirajan and Gaur, 1997). Linschoten (1903) who stayed in Goa from 1583 to 1588 observed that 'ships of 200 tonnes or little more may easily discharge before the town of Goa upwards into the river'. Ships were hauled ashore during rains and thatched to preserve the decks from dampness. This indicates the fall in commercial activity during the rainy season. From the beginning of the 17th century Old Goa started losing its importance as a port and commercial centre. The struggle between the Dutch and the Portuguese continued for sixty nine years. Slowly Portugal lost her flourishing commerce. Epidemics, poverty and violence ruined the Portuguese (Fonseca, 1986). The regular blockades by the Dutch disrupted maritime trade so much so that between 1641 and 1644 not a single carrack could sail to Portugal from Goa (De Souza, 1979). Tavernier, the French traveller who visited Goa between 1642 and 1648, refers to poverty in Old Goa, which he attributes to the incursions of the Dutch and the English.

Shipbuilding during Portuguese Rule

Until recently, the only evidence for ancient boatbuilding in Egypt, Mesopotamia, the Indus Valley and the Persian Gulf region came from references in cuneiform texts, models from various excavations and representations on seals, sealings, potsherds and temple art or petroglyphs (Winkler, 1938; Bowen, 1956; Bass, 1972; Qualls, 1985; and Rogers, 1992). In the Indian context information on the use of timber in the shipbuilding industry during the historical period comes from the writings of Panini (5th century BC) (Agrawal, 1953), Patanjali (2nd century BC), Strabo (60 BC-19 AD) and Arrian (200 AD). The *Yuktikalpataru* (The Wishing tree of Artifice) compiled by King Bhoja of Dhar (11th century AD) recounts the timber used in the construction of ships. It mentions that different varieties of timber used in different parts of a ship. Indian teak was preferred for shipbuilding because of certain inherent qualities, such as light weight combined with good strength; availability; resistance to corrosion by sea water and attack by marine organisms; and good retention of shape and size under various conditions of wetting and drying. In addition timbers used for shipbuilding needed to have resistance to electro-chemical breakdown and be non-corrosive to metals. Indian teak possessed all these qualities. Al Masudi (871-957) and Ibn Jubayr (1145-1217) mention the use of Indian teak in shipbuilding. It was commonly used and hence was in demand in Middle Eastern countries and later by Europeans. Europeans replaced oak and pine with Indian teak in shipbuilding industry because of its versatility and high quality.

The Portuguese desired to dominate as much of the East as possible and have undisputed control over trade, commerce and navigation. Ever since their arrival on Indian littoral and throughout the 16th century, they had made attempts to capture and control strategic ports and trade centres on its west coast. Over a period of time they constructed twenty one forts and established a network of strong trading posts all along the west coast (Mathew, 1982). One of the dreams of Afonso de Albuquerque was to seize the shipbuilding industry of Goa, which was one of the best centres in India during that period (Danvers, 1988). Francisco de Almeida, the Portuguese viceroy, sent a letter to the King of Portugal advising on the importance and requirement of a strong naval fleet at the ports, trade centres and trade routes. Almeida wrote the following to King Manuel I 'Let it be known for certain that as long as you may be powerful at sea you will hold India as yours and if you do not possess this power little will avail you a fortress on the shore' (Danvers, 1988). In one of letters to the viceroy, the King of Portugal emphasized the importance of building ships in India. The royal letter reads, 'Many occasions have uncovered the requirement of the State to have *galleons* and *naus* of high freeboard and I have received information that they are always kept over there and without them, it is not possible to retain control over the seas, in which lies all power of this State. For this reason, I would like to recommend to you and trust you, that despite all difficulties that may come your way, you shall try each year to have at least four *galleons* built, of suitable tonnage required for these regions. Their building shall be laid out at a place where timber is available in great abundance and they can be built with great convenience' (Cunha Rivara, 1857-1877). Initially, the Portuguese decided to develop the existing shipbuilding yards of Bassein, Cochin, Daman and Goa and ten ships of *carreira* were built. As early as 1512 the Portuguese had 50 ships in India and the number had rapidly increased. By 1525 there were varied types of ships in Portuguese India, namely, 11 *galleons*, 9 lateen ships, 6 *naus* and barges (Mathew, 1988). Moreover, ships were sent to Macau, Mombasa, Timor and other countries.

Timber and Shipbuilding

India has contributed to world culture in several fields including shipbuilding technology. Recent research illustrates that Indian teak was preferred for shipbuilding by Indian and other mariners due to innate qualities. Before coming to India, the Portuguese used oak and pine in shipbuilding. Ships built of these woods were frequently affected by teredo and other marine organisms. The maintenance cost of the ships was relatively high compared with their longevity. Oak contain a powerful organic acid which corrodes iron fastenings. It is stated that the life of a ship exceeds 50 years if is made of teak, whereas it lasts for 12 years if made of oak (Roy, 1972). After recapturing Goa on 25 November 1510 Afonso de Albuquerque found a considerable amount of arms and ammunition along with naval munitions in the dockyard of Old Goa. The building of ships continued immediately after the conquest and he initiated measures for the maintenance of warehouses, armoury and stables. Albuquerque wrote a letter to the Crown in 1510 stating the availability of expertise and excellence of Indian carpenters. Diniz Fernades de Melo was nominated as the Chief Master of the ships; and the King took keen interest in seeing that all these establishments were well managed. A royal order of 1585 to Viceroy D. Duarte de Menezes and again in 1595 to Viceroy D. Mathias de Albuquerque emphasised the importance of construction of *carracks* for the *carreira* in India but not in Lisbon because the availability of quality raw material and expertise confirmed that Indian-built ships last much longer than Lisbon-built ships. Furthermore Indian-built ships were cheaper and stronger (Boxer, 1969 and 1980). By the Royal decree of 22 February 1613 revenue amounting to 60,000 *xerafins* every year was to be allocated to the dockyard and no portion of this amount was to be used for any other purpose. It was used exclusively for repairing and rigging the ships which sailed from Lisbon. Further, it was advised that no payments for timber or nails should be made in Lisbon as the price was high. In 1612, an order was issued to buy wood for shipbuilding from the north of Goa as it was cheaper over there. The viceroy complying with this royal advice was informed that a contract for building ships in the north had already been signed.

Thus the Portuguese grasped the superiority of teak which was plentiful in the hinterland and used extensively in the shipbuilding industry (D'Silva, 1997; and Mathew, 1997). All Europeans borrowed ideas in various fields from India. The Portuguese also adopted measures to regulate the timber trade in India. In addition to teak, the other timber used in shipbuilding was *angeli* (*anjili*) wood, which was plentiful along Malabar and other parts of the west coast of India (Mathew, 1988). Like teak, cordage and cables were highly valued because of their strength and durability and to make them stronger they were kept in salt water (Qaiser, 1982). The *Cartaz*, a sailing permit issued by the Portuguese to private merchants, clearly prohibited them from carrying teak on board to any port (Martins, 2008). Subsequently, the Portuguese promulgated orders prohibiting the buying and selling of teak for any purpose. Even the felling of trees in Bassein Forest was prohibited without the prior permission, except for the repair of houses. Further, contractors to whom forest leases were granted had to obtain a permit for cutting down trees to supply the dockyard and they were fined for violations. Of the total amount of the penalty collected, half was given to the informer and the remaining half was spent on the hospitals. No tax was paid by the Portuguese on timber (Pissurlencar, 1857-1877). Later some of the local rulers on the west coast supplied timber and other materials to the Portuguese for shipbuilding. In one of his letters Afonso de Albuquerque informed King Manuel of Portugal that, 'the King Zamorin is willing to supply timber for the construction of *naus*, *galleys* and other types of ships and the ruler of Amin Island has signed a treaty to supply coir (coconut fibre) required for shipbuilding' (Mathew, 1988). The Portuguese obtained supplies of timber from Bassein, Cannanore, Cochin and Kanara. Even Portuguese *fidalgos* who owned forests around Bassein exported timber to Goa for shipbuilding (David, 1957). Indian timber became so much in demand that in the later period, Portuguese rulers in India signed treaties with the Raja of Cochin and Adil Shah of Bijapur to supply timber for ship construction. The ruler of Ikkeri Virabhadra also agreed in the treaty of 1633 to provide timber to the Portuguese for the same purpose. Apart from acquiring shipbuilding timber, the Portuguese obtained masts and yards from Cochin for their *galleons*. Coir was exported from Goa and teak was even exported to Brazil for the construction and repair of ships (Martins, 1994).

Proper treatment was given to timber at the shipyards. Before use timber was immersed in proximity to the river mouth, where the water was mixed with sea water. This treatment enhanced its longevity and inhibits damage by teredo. Sometimes timber was buried in the sand to conserve humidity and also kept in sea water, because prolonged exposure to air could result in loss of resin and diminish resistance. Great care was taken when stacking timber in the warehouse. The floor of the warehouse was kept dry and a sufficient gap was created between stacked timbers. In the case of low warehouse roofs, ventilation holes were provided in the walls. Quantities of cordage were imported from the Maldives, iron nails were imported from Biscay because it was cheaper and the quality was better as compared with that of Lisbon. Copper was brought from Europe for the purpose (Martins, 2008). After the Portuguese initiated the building of their ships in India, a number of changes in design and construction took place such as the introduction of nails as fastenings and an increased use of caulking in joints (Deloche, 1996). The Portuguese borrowed wood joinery technology from the traditional shipbuilders of India. Indian carpenters worked under the European chief-carpenter's supervision and they were exposed to the European technique of caulking (Qaisar, 1982). Afterwards, Indian shipwrights incorporated the techniques applied by the Europeans in native designs (Arasaratnam, 1994). Indian ships were stitched with coir to join planking and it was a perfect technique but unable to take the impact of artillery. Indian shipbuilders realised that European vessels fastened with iron were stronger in such situations. Portuguese observed that teak and other hard wood varieties of timber such as mango, *matti* and *angeli* were suitable for shipbuilding and they studied and compared with the European varieties of oak and pine (Martin, 1994). The other timbers used were *Ash, Buch, Guaiac, Acacia, Catchu, Mahogany, Brazil wood* and *Chestnut*, but these timbers were not resistant to a marine environment. Furthermore cables and ropes were made of hemp. Utmost care was taken to use high-quality material. No defect and damaged timber was purchased. The wood for construction and repair had to be stacked separately according to cost, quality and proportion. The quality of material was examined and identified by competent authorities (Shastry, 1981). During the 15th and 16th centuries Portugal was the only European country where economic, military and technical factors favoured the conception of the round ship.

Shipwrights from Portugal visited India and noticed the superiority of Indian timber, recommended its use and suggested the same for the Portuguese ships instead of pine and oak. During the sixteenth century many Portuguese *naus* were built of Indian teak (Stephen, 1997). The extensive use of timber resulted in a shortage of teak in Cochin and other places along the west coast of India (Scammell, 1995). Similarly, coir used for stitching and making cordage was an item of merchandise for the Portuguese (Panikkar, 1997). Over a period of two centuries, tactics in naval warfare led to developments in hull design, sailing rig and mounting of heavy guns. Developments in the shipbuilding industry were not a simple historical evolution; it was a long process to replace inferior old ships with new types of superior quality. Precise information was obtained about the required tonnage before construction. Ships were designed to meet the technical needs of the long voyage on the high seas. Similarly, the dimension of each part of the ship depended on the size and tonnage of the vessel. The timber used for construction of seagoing ships was of better quality so that it could be shaped and curved as necessary. It also provided stability and sturdiness to resist the fury of the seas, cyclones and storms. Seasoned timber would not perish in a marine environment and protected the attacks of *Busano* (shipworm) (Barata, 1989). In the 15th century the first phase of development had taken place when three-masted square rigs were introduced. In the latter part of the century square rigs were retained forward and a triangular lateen sail raised on a mizzen mast. The main mast continued to carry most of the canvass to drive the ship but the sail area could be spread forward and aft to balance the ship on any point of sailing, and by adjusting the rig the ship could sail on any heading. In this way large ships could be handled in conditions in which even small ships with a single mast and sail had been helpless. Thus there was an enormous advantage in both economic and military terms. Ship plans and designs developed to such an extent that it became easy to differentiate between warships and merchantman. The installation of artillery on board Portuguese vessels was mostly in proportion to their tonnage. Areas of operation were controlled by geopolitics. Portuguese and other European powers challenged each another for a toehold in the Indian Ocean. The Portuguese rigged and equipped '*galleons*', '*naus*' and '*frigates*'. Keeping in mind the needs of the eastern voyage and warfare during the early days of their voyages, the captains of ships were instructed to carry naval spares to carry out repairs during voyages. Every European power began to arm ships with heavy cannon. The tonnage of ships increased as armament and a greater volume of cargo were loaded and the movement of people increased.

The shipbuilding centres of Portuguese India were at Old Goa, Bassein (Maharashtra), Cochin (Kerala) and Daman on the western coast of India both for construction and repair of ships. During the Portuguese regime the dockyard at Goa was highly organised (Boxer, 1959). There were three dockyards and shipyards between the city of Goa and the riverbank of the Mandovi. The dockyards of Old Goa had several names, namely, *Rebeira, Rebeira das Armadas, Rebeira das Naus* and *Arsenal de Rebeira da Naus de Cidade de Goa*. The superintendent was entitled Feitor (Fonseca, 1986). The naval dockyard at Old Goa was the key Portuguese naval establishment in the whole of *Estado da India*. If any ship was to be purchased by the Portuguese either at Bassein or Daman, technical experts from Goa were called to make the inspection and valuation of the ship. Pyrard de Laval, who visited Goa gives the ensuing statement, 'There were three dockyards between the city of Goa and river bank, the main dockyard was on the western side of the city was known as *Ribeira das Naus* or *Ribeira de Grande*. Gun foundry and mint were present in this dockyard. *Cais de Santa Catarina* was next to the dockyard. This remains busy every time Armadas were anchored. The other annexe of the dockyard was known as *Ribeira das Gales*, and this was fortified' (Laval, 1601-1611). After unloading the cargo, ships moved towards *Cais da Fortaleza do Visorey* which was located between the palace of the viceroy and the river where all the Indiamen anchored. As regards naval construction in the 16th and 17th centuries at the Royal Dockyard a number of ships were built, among them *Santo António de Tanná, nau Cinco Chagas, Nau São João Baptista* and *Nau São Francisco, galleons Bom Jesus* and *Madre de Deus* (Hunter, 1899). *Nau Cinco Chagas* served for 25 years and made 9 or 10 trips, excluding other voyages, before ending her services (Mathew, 1988). The carrack *Cinco Chagas* was built under the personal supervision of Viceroy Dom Constatino de Braganza, who selected timbers piece by piece. Highly experienced and knowledgeable naval experts were appointed at the dockyards. Amongst them António Gonsalves, Master Caulker worked for 26 years in the royal service and during his time the cost of shipbuilding was reduced. In order to meet Dutch challenges António da Costa was appointed Master Caulker who was able to equip the galleons responsively (Martins, 1994). The masters of the shipyard during the Portuguese period were Diogo Luis who built a *carrack São Joao Baptista* in 1621 and a great *galleon Bom Jesus* of 1200 tons, which was launched in Goa in 1636, with 70 copper pieces (artillery), described by a German traveller as one of the noblest ships that he had ever seen. Similarly, repairs to a number of ships were undertaken. Early Portuguese ships were of lesser tonnage and moderately equipped, but tonnage and size increased after the discovery of the sea-route between India and Portugal. The ships that plied it usually had a capacity of 400 tons. After 1570 the tonnage of ships exceeded 600 tons. Under Viceroy Conde de Linhares the shipyard of Goa came under close scrutiny and he visited regularly to inspect the construction of ships until they sailed to Lisbon (Disney, 1984). More *naus* and *gales* were used in the 14th century, but in the 15th century progress was made in the construction of *caravelas*. The variety of ships can be inferred from Portuguese documents which relate to maritime trade. For example: *Albetoca, Almaida* (80 ft long and 6-7 ft broad), *Barca* 10 to 30 tons, *Canoa* (large vessel of single piece carved out), *Carraca* (could carry 2000 people), Caravela (below 200 tons), *Fusta* (300 tons), *Gale* (25 to 30 oars on each side), *Galeão* (100 to 1000 tons), *Nau* (100 to 120 tons) and *Pinaça*. Some of the vessels were used on sea voyages and some for inland trade, carried heavy guns, artillery for warfare, cargo and horses. It is known that Vasco da Gama used *caravelas* during his voyage to India.

Shipping

When Vasco da Gama's *naus* dropped anchor in the placid waters of Calicut on 21 May 1498, it was an event that brought about tremendous changes in the history of Portugal, Europe, India and the Far East. For the next one or two centuries, Europe and Asia were destined to come in contact with each other on all fronts commercial, political, scientific and socio-religious undreamed of by any visionary. The era of Vasco da Gama had been ushered in. Prior to this, Portugal, having freed herself from the Moorish wars of the medieval period, had imbibed the spirit of the Italian Renaissance ready to explore anything that was new or unknown. By the beginning of the fifteenth century, the Portuguese were not only important traders on North European sea-routes, but had begun to look southwards, discovering and annexing the Canary Islands and exploring the west coast of Africa. By 1470, they had successfully examined and mapped a large portion of the African coast. Their spectacular success when Vasco da Gama reached Calicut for spices and Christians would have hardly been possible without the organising genius of Prince Henry, the Navigator. He had himself fought as a crusader in North Africa and was the first to channel the energies of the enterprising Portuguese merchants and *fidalgos* into discovering a new sea-route to the Old World, i.e. India, and the countries beyond, were known for their spices and fabulous wealth. Prince Henry surrounded himself with geographers, cartographers, navigators and ship designers. One of his most notable achievements was perfecting a new type of ship, the *caravela*, able to brave the storms of the high seas without having to hug to the coast. The Mediterranean *galleon* depending upon oars had become out of date for oceanic traffic. Henry realised this. He devised for his expeditions a three-masted *caravel* of 50 to 200 tons, which admirably suited the deep and distant sea voyages. Henry financed expeditions along the west coast of Africa, Madeira, the Azores, Cabo Verde and Cabo Bojador. Trading was made profitable by gold and slaves secured at these places. Prince Henry established the naval school at Sagres and did splendid work in shipbuilding and nautical science. He died in 1460, before completing his work. His successors under changed circumstances could not maintain the high ideal of the Navigator. The *naus* of Henry made Vasco da Gama 'Admiral of the Indian Ocean', and the King of Portugal, 'Lord of the Conquest, Navigation, and commerce of Ethiopia, Persia and India'. King John succeeded Afonso V to the throne of Portugal. He took up the projects left untouched since the death of Prince Henry. But towards the end of the 15th century all the glory of the Portuguese echoed by these high-sounding titles had faded away and in the following centuries Portugal continued to exist as a maritime power only because she was tolerated by the powerful Dutch, English and Spanish rule (Weech, 1961).

In the early part Portuguese were the masters of the seas and their navy was strong and dreaded by their enemies. Portuguese desired to acquire products from the African coast and crush the Arab power and trade to discover the extent of the Muslim territory and spread Christianity were also some of their driving motives besides their love for adventure. The commercial motives were that exotic spices and aromatic goods spices, pepper, cardamom, indigo, saffron, pearls, ivory, silk, muslin, carpets, fetched good revenue. The profits derived from customs duties were sent to Goa for general state needs. Therefore Goa can be viewed as the centre of Portugal's seaborne empire getting revenue not from its own economic activity so much as its dominance in Asian waters. Portuguese mariners were aware of the monsoon winds. An experienced pilot of the *carreira da India*, when asked what was the best season for departure of the annual Indiamen from Lisbon is alleged to have replied: 'The last day of February is time enough, but the first day of March is late'. In practice, the ships often left in the second half of March or in the first half of April and belated departures in May were not unknown. The Southwest monsoon which normally begins on the west coast at the beginning of June had the effect of virtually closing all harbours in this region from the end of May to early September. Outward bound Indiamen therefore aimed to leave Lisbon in time to round the Cape of Good Hope in July and to reach Goa in September or October.

Life on Portuguese ship

It was the dream of every Portuguese to make a voyage to India, but the majorities were not accustomed to the sea conditions of the Indian Ocean. They faced many hazards, storms and uncertainties. Regarding a voyage to India, an anonymous document mentions that the voyage is so hard that you will run out of tears, before you run out of reasons for shedding them (Disney, 1990). The ships coming from Lisbon had artisans and craftsmen of all kinds on board for necessary repairs, such as carpenters, caulkers and coopers. The space allotted for each passenger was fixed by law, whereas *fidalgos* and other nobles were given privileges. The lack of space on board compelled many to sleep on deck exposed to rain and sun. Most of the cabin boys were attached to passengers. Four boys were expected to sleep in the crow's nest (*cesto da gãvea*). The master, boatswain, guardian or quartermaster and the chief of the gunners had their own silver whistles hanging from a silver chain on the neck to summon men for specific work namely the master and the boatswain to call the mariners; the chief of the gunners to call all the gunners and the guardian to call the cabin boys and pages (Mathew, 2002). During voyages to India, Portuguese ships experienced a deficiency of food. The quantity of stores and supply of food loaded on board was not always proportionate to the numbers of passengers. The means of storing food and other eatables was in unhygienic conditions. Alexander Valignano, the Jesuit missionary priest, has furnished detailed information (Boxer, 1969). The unavailability of fresh water, food, sanitation and the lengthy voyage caused different diseases, such as beriberi, scurvy and fever. Unhealthy conditions caused a high mortality and this led to the spread of beriberi, scurvy and other disease (Boxer, 1960).

A ship arriving in India from Lisbon during the 16th century was a microcosm in which personnel of different ranks and professions voyaged together. During the voyage each had a special duty to perform. The social composition of the people on board was unusual. It did not lack the tensions found in any other society and they were perhaps more evident due to the environment. People were always alert to the fact of death and so some of them took a positive step towards this universal phenomenon while others cherished a negative attitude. The numerous religious practices actively participated in by many speak for the anguish of death and a desire to be prepared for it. The discipline on board, to a large extent, depended on the religious. Rosaries were said on every Monday, Wednesday and Friday evening. Charitable works performed by the religious were much appreciated by the sick. One may say that life on board a Portuguese ship during the sixteenth century differed from that on indigenous ships due to the duration of voyages and the social background of the people (Mathew, 2002). The usual amusements and entertainments were allowed during stipulated times and beyond that were punishable (Scammell, 1982). In order to avoid potential dangers to the ship and passengers the ancient antiphony was sung in honour of the Virgin Mary.

Shipwrecks

The coast of Goa is evidently dotted with many shipwrecks. Records of pre-Portuguese period shipwrecks in Goa waters are very scanty and available information on shipwrecks off Goa coast date from the Portuguese period onwards. Only one reference to a pre-Portuguese shipwreck is available and it dates to the early medieval period. The inscription of Jayakeshi-I (1052) of the Kadamba dynasty points out that it occurred while Guhalladeva-I (980-1005) was sailing on a pilgrimage trip to Somnath, on the Saurashtra coast of Gujarat, in a ship from Chandrapura (modern Chandor). He had reached half way when the mast of his ship broke and he took shelter at the nearby port of Goa probably the port of Gopakapatana. The stranded king was helped by a wealthy Arab merchant named Madhumat (or Mohammed) of Taji origin, who was the chief of the Arab trading (*Hanjamananagar*) settlement in Goa (Gune, 1979; and Moraes, 1990). Further, Mascarenhas (1987) mentioned that a Kadamba period ship was wrecked near St. Joao Island (off Gopakapatana) at the mouth of the River Zuari. However, Mascarenhas did not supply other details of the shipwreck. It appears that his reference was derived from the inscription of Guhalladeva-I. The *hero stones* depicting naval battle scenes, now housed in the Archaeological Museum (ASI) in Old Goa, confirm maritime activities in Goa during the Kadamba period (12th-13th centuries AD). Portuguese marine, shipping, trade and commerce records housed in the India House, Lisbon and Goa State Archives, Panaji mention that a number of ships were wrecked in Goa waters during their voyages to and from Portugal (Mathew, 1988). Some records dealing with the *carreira* which were stored in India House were destroyed in the Lisbon earthquake of 1755, but Boxer (1957) compiled shipwreck disasters from the records of '*Casa da India*'.

The new type of *caravel* enabled Portuguese mariners to cross oceans. But when *caravels* were neglected, shipwrecks resulted, bringing immense misery to every home in Portugal. Everyone, from a sailor to a captain, and from a *fidalgo* to a *frei*, was involved in shipwrecks. A peep into the records enables us to understand the causes of shipwrecks, which among other reasons were responsible for the decline of the Portuguese navy. Between 1497 and 1612, 806 ships sailed to India, but only 425 returned to Portugal and as many as 20 ships ran aground, 66 were lost in shipwrecks, 4 were captured by the enemy, 6 caught fire and the rest (285) remained in India. Sub-divisions of this period show how shipwreck rates changed. The proportion of ships wrecked during the 16th century was alarming. The average cost of a ship was 125,000 *cruzados*. Calculated at this rate the cost of the 66 ships which were wrecked and the 20 ships which ran aground shows that the Portuguese suffered a tremendous monetary loss of 125,000 X 86 *cruzados*. In the beginning the *cruzado* was made of gold. Later, it was silver and worth Rs 4 to 4.80. The majority of ships were wrecked in the Indian Ocean and most of the tragedies happened on the return voyage off the coast of Natal before the ships rounded the Cape of Good Hope. The tragedies were much less once the ships crossed the Cape (*Cabo da Boa Esperanca*) (Boxer, 1957; and Fonseca 1986). On the other hand storms and cyclones appeared which could not be predicted, because mariners lack meteorological knowledge. As the equatorial region was very dangerous changing winds and currents were responsible for many shipwrecks. Sometimes the current in this region was fatal and was called the '*Vara de Coromendel*' (Mathew, 1988).

The primary causes of loss of ships and people continued due to overloading on the homeward run, combined with sailing too late in the season from either Lisbon or Goa. Ships which left Goa too late ran into stormy weather off the Southeast African coast. This was particularly true of the lubberly four-deck *carracks* which were so much in vogue in the *carreira* during the years 1580-1650 and whose structural defects were aggravated by inefficient careening and by reckless overloading at Goa (Boxer, 1959). During the 15th-

17th centuries every maritime nation had sufficiently experienced seamen for overseas sailing. The English were largely dependent on landsmen secured by the pressgang and the Dutch Navy had a significant proportion of Scandinavian and German seamen, as did the ships of the Dutch East and West India Companies. In comparison with other European countries the Portuguese had very few harbours along their maritime trade-route and there were no large seamen such as the Basque and Cantabrian ports often provided for Spain. The *carreira* was therefore constantly struggling to get good seamen. Sometimes men were paid to be entered as sailors aboard Portuguese Indiamen; but the rest of the time they were forcibly embarked and kept on board in chains, to prevent desertion before the ship weighed anchor in the Tagus (Boxer, 1959).

V. Magalhães Godinho has made a meticulous comparison of the surviving records and states that between 1500 and 1635 about 912 ships sailed to the East, but only 768 reached their destination, whilst on the homeward voyage 550 ships sailed, but only 470 arrived safely (Boxer, 1969). The study of a variety of marine records shows that the majority of Portuguese ships were wrecked in the shallow waters off Goa due mainly to uneven seabed topography, hidden reefs, sandbars, severe storms and lack of proper maintenance. However, exact locations cannot be derived from these records. Descriptions of a few wreck events enable us to know when a ship was actually overwhelmed by a storm. Accounts of shipwrecks were left by survivors. Afonso de Albuquerque and the epic poet of Portugal, Luis de Camas, were themselves shipwrecked (Blackmore, 2002).

The most common factor was overcrowding of ships bound for India and Malacca and the return voyage to Portugal. Frei João de Lucena, the author of *Vida de Sam Francisco Xavier*, described the overcrowding on a Portuguese ship bound for India and wrote, 'Imagine for yourself and think well what a ship of India is like when put to sea, with six hundred, eight hundred and sometimes a thousand persons in it comprising men, women, children, slaves, *fidalgos*, merchants, soldiers and sailors. Five months was essential for a voyage to India and back. During the voyage one had to face many dangers; the doldrums of the Guinea Coast, rotting of provisions and stores, lack of accommodation, fever, sea sickness, spells of darkness and constant dangers of death' (Mathew, 1988). In this condition shipwreck was not any worse for the unprivileged and was perhaps a boon for slaves. Further, ships returning to Portugal were packed to capacity with spices, cloth and other valuable commodities. Everyone from the captain to the cabin boy packed the cargo space allotted to him according to his status, known as the 'liberty chest', in order to sell the contents on returning home. This was one of the things that tempted men to seek service in the Portuguese navy. The other problem was trimming the ships. Generally, heavy goods had to be put in the lower part of the hull and lighter articles were to be piled on top. In Portuguese ships it was reversed, because it was easier to unload heavy goods if they were kept at the top. This arrangement made the ship top heavy and disturbed the entire balance of the ship. Ships were neither regularly repaired nor properly inspected by the authorities (Boxer, 1960). This was particularly the case when repair was undertaken under contract. Careening caulking, refitting timber, stitching the sails, nailing and a hundred other jobs were finished as quickly as possible to save time and labour. Even inspecting officers could be bribed. All these problems and defects were exposed when the ship faced a storm and remedy was too late. Very often old ships put to sea for long voyages which were not in a position to withstand ocean conditions. This was inevitable as Portugal was short of ships to maintain her seaborne empire. In fact she did not possess more than 300 ships when her maritime glory was at its height in 1536.

There was an episode when after the conquest of Malacca in 1511 Afonso de Albuquerque was returning to Goa with a rich booty and his flagship '*Flor-de-la-Mar*' (Flower of the Sea) was wrecked near Sumatra (Mustapa, 2005). Albuquerque and his companions swam ashore. The ship was wrecked because it was old and leaky. It was carrying a table of gold studded with jewels on which the Queen of Malacca took her meals. The table was valued at 800,000 *cruzados*, and the precious stones at 3000,000 *cruzados*. The trade between India and Portugal continued until the mid 20th century and during this period several ships were wrecked in along the Indian waters. The National Institute of Oceanography (NIO) has collected shipwreck data from marine records off the Goa coast (Gudigar et al., 1995-96). In addition Portuguese records mention Portuguese shipwrecks off Colombo, Gulf of Ceylon, Socotra and the Sumatran coast. Furthermore, shipwrecks off the Goa coast are marked on Hydrographic Chart No. 2020 and local divers, fishermen and scholars were consulted for information on shipwrecks. The marine records denote that an unseasonable hurricane which struck the River Mandovi in April 1648 sank twelve Portuguese ships en route to Calcutta (Kolkata) near Aguada Bay (Boxer, 1960; Gudigar and Sundaresh, 1992; and Thakkar, 1991a-b). Similarly, in 1651, five Portuguese ships, namely, *Santa Helena* (1654-55), *São João Evangelista* (1652-53), *St. Francisco* and *St. Thome* set sail from Portugal, but only two ships reached the coast of Goa to be wrecked in a storm on 10th March 1651. Captain Joao da Costa was the captain of *São João Evangelista* (1652-53); and Captain Manual da Pina da Cunha of the *Santa Helena* (1654-55). Both ships were wrecked on the bar of Goa. A Portuguese ship, *Pataxo Santa Tereza de Jesus*, was ready to sail to Bassein and Chaul in Maharashtra but was wrecked off the sandbar of Goa on 18 May 1658. Some cargo was salvaged and the other remains were looted. A small bag containing precious stones, royal money and other goods on board the *Manoel Dias* went down in the sea (Chauhan, 1988). The *St. Cristovão* and *Santo André*, both Portuguese sailing vessels, were wrecked in natural calamities. The majority of records indicate that a number of ships were wrecked off the sandbar of Aguada Bay (Gudigar et al., 1995-96). Some of the mentions of shipwrecks are cited below: One record details the tragic story of the *Brig Champion*, an English East India Company ship lost in Goa waters. The ship sailed from Bombay to Karachi on 30 August 1862. Two days later Newman, the captain, fell severely ill. The second officer of the ship, Alexander Fraser, was made first officer and the voyage continued. On 7 September 1862 at Lat 10^0 43' N and Long 69^0 22'E the sea became stormy and choppy; the crew refused to work the pump unless the course of the vessel was changed in order to return to Bombay. As the Captain was very sick, the first officer ordered an immediate change of course to land whatever possible. The vessel reached Aguada on 13 September 1862 and immediately Newman was sent to the onshore hospital where the physicians pronounced his disease incurable. On the evening of 15 September 1862 the *Brig Champion* left the port of Nova Goa and while crossing the Aguada sand bar the vessel refused to obey the rudder and lay entirely at the mercy of the wind. All efforts to take her by any course whatever having proved futile, the pilot was ordered to anchor the vessel which was at once done, the vessel at the same time grounding. The first officer slipped about 8 to 9 fathoms of cable. In the meantime the pilot lost his nerve and commenced to pray without caring for the safety of the vessel, all his attention and energy being apparently directed to laying hold of the stays of the mainmast and calling upon the mercy of God to save him. Viewing the poor old man in this distressed state and having in the meantime a slight improvement in the weather, the first officer attempted to weigh anchor in order to extricate the vessel from its position, but all in vain for the vessel continued with redoubled strength to strike on the bar. At 1 AM while the side of the vessel was giving way all on board jumped overboard. Crew and passengers finally succeeded by means of a small boat in reaching the shore, losing only one individual. On 16 September Alexander Fraser could see nothing of the wrecked *Brig Champion* except broken timber and planks floating towards the shore (*Marine Records Vol. 1862*).

Other Marine Records namely '*Ministẽrio Do Ultramar, Arquivo Histõrico Ultramarino, India of 30 Marc̱o 1695 and 15 Junho 1696*', remark the following information on the shipwrecks off Goa port: "In 1694, *Nossa Senhora do Rosãrio* went aground on its voyage to Goa port. After having given to your Majesty a detailed account of the success that your fleet of ships had in India and on the recommendation that I had suggested for the strait. It happened that two ships ran aground. One is being *Nao Sra de Vale* which was salvaged without any damage except losing propeller and a little water leakage into the ship. The other ship *Nao Sra do Rosãrio* got stuck between the rocks. We tried our level best to salvage whatever we could especially the artillery, which was soaked in water and parts of the ship to make a new one. I was not aboard the ship for some time as I was sick. In the expedition the troops of the regiment and I do feel guilty and I blame myself, as the master of the frigate *Rosãrio Dom Vasco* thanked the Lord for all the good done in salvaging the vessel. I blame this incident for my misfortune. And I believe that this is not going to affect the prestige of the nation but only the convenience of sending our flotilla through these straits. I have instructed that all the sailors from these two frigates should be put aboard the frigate *Nossa Senhora de Conceição* which was well provided and stocked. The frigate *Santo Francisco Xavier* and a battleship were dispatched to the straits which had good luck. In the meanwhile after having written to Your Majesty we got news of the invasion of Arabia". Recently, the Centro Nacional de Arqueologia Náutica e Subaquática (CNANS) in Lisbon has compiled a list of Portuguese shipwrecks in the waters off Goa and along the Indian coast using archival and archaeological sources (Alves et al., 2007). These records provide information on wrecks but cannot specify their exact locations. The Portuguese shipwrecks recorded in the marine records are detailed in Table - 1.

S.No	Name of the ship	Year of wreck	Location	Destination	Remarks
1	*Sao Jorge*	1510	Padua Reefs near Cananor	Return voyage	The ship was grounded
2	*Sao Vicente*	1510?	Laquedives		Grounded
3	*Nossa Senhora da Nazare*	1522	Bar of Goa	Coming to Goa port	Cap. D. Pedro de Castro, wrecked due to storm, ship was old
4	*Esperanca*	1531	Calecare, near Cape Camorin	Outward voyage	-----
5	*Vera Cruz*	1532	Nicobar Islands	Return voyage	-----
6	*Santiago*	1542	Bacaim, modern Vasai and Versava	Outward voyage	While sailing to Goa in 1542 met with a furious storm and as a result she was reduced into pieces in the river 'Cabras' between Bacaim, and Versava.
7	*Espirito Santo*	1543?	Malabar coast	Return voyage	-----
8	*Frol de La Mar*	1551	Goa (40 leagues from) or near Bacaim	Outward voyage	The ship wrecked due to storm.
9	*Santa Cruz 'Zambuco'*	1552	Seitapor river, near Goa	Outward voyage	Set ashore
10	*Sao Boaventura*	1554	Bar of Goa	Coming to Goa port	Cap. D. Pedro de Mascarenhas, wrecked due to storm
11	*Sao Filipe*	Sep 1563	Bar of Goa	Coming to Goa port	Cap. Vasco Lourenco de Barbuda, due to storm, all cargo went into the sea.

Contd...

12	*Reis Magos*	20 Apr 1582	Cochin	Portugal to Cochin	Wooden sailing ship, on return journey the ship wrecked due to natural calamity.
13	*Nossa Senhora das Reliquias*	5 Jan 1587	Bar of Goa	Coming to Goa port	Cap. Francisco Carvalho, due to excess of cargo
14	*Barco Portugues*	April 1594	Ao largo de Goa	De China para Goa, India	The ship was carrying Ouro e porcelana.
15	*S. Cristovam*	17 Aug 1594	Off Goa waters	Portugal to Goa	Wooden sailing ship; sunk when she came back to Goa due to natural calamity
16	*Religuies*	1594	Bank of Cochin	Lisbon to Cochin	Wooden sailing ship; she left Lisbon on 11 Apr 1586 and sunk on 18 Aug 1594.
17	*Madre de Deus de Guadalupe*	6 Jan 1598	Cochim bar	-----	-----
18	*Madre de Deus*	30 April 1598	Cochim	Return voyage	-----
19	*Nossa Senhora de Betancor*	1606	Goa	Return voyage	-----
20	*Nossa Senhora do Loreto*	20 Oct 1607	Wrecked in the burnt island of (Queimados)	-----	Cap. Jeronimo Teles de Albuquerque, ship was fired by Dutch.
21	*Nossa Senhora dos Loretos* or *do Loreto*	1607	Queimados Islands, Goa	Outward voyage	Burnt by the Dutch
22	*Santo Andre*	27 May 1608	Ao largo de Goa - Bar of Goa	De Portugal para Goa	Cap. Luis de Brito de Melo; loss due to natural calamity; wooden sailing 'Nau' ship. It was also carrying Moeda.
23	*Nossa Senhora da Oliveira*	1608-09	Queimados Islands, Goa	Outward voyage	The ship was burnt
24	*Nossa Senhora da Nazare*	1615	Surrate (Surate)	Outward voyage	Burnt in battle against the English
25	*Nossa Senhora dos Remedios*	28 Jan 1616	Bar of Goa	Coming to Goa port	Cap. Paulo Rangel de Castelo Branco, dur to storm, crew saved and cargo onboard lost. Wooden sailing 'Nau' ship.
26	*Madre De Deus*	4 Aug 1635	Panaji, India	De Portugal para a India	The ship was carrying Moeda.
27	*Sao Joao de Deus*	1 Mar 1638	Bar of Goa	Coming to Goa port	Cap. Goncalo de Barros da Silva, due to fire.
28	*Madre De Deus*	1638	Goa, India	De Portugal para a India	The ship was carrying Moeda

Contd...

29	*Nossa Senhora da Quietacao*	22 Sep 1641	Nearer to Goa	Coming to Goa port	Cap. Sancho de Faria da Silva, due to naval battle with the Dutch and sunk in battle
30	*Sao Joao Evangelista*	1652	Bar of Goa	-----	-----
31	*Santa Helena*	1654	Aguada and Mandovi bar, Goa	Outward voyage	-----
32	*Santa Terresa de Jesus*	15 May 1658	Mandovi bar, Goa	Return voyage	-----
33	*Santissimo Sacramento*	Sep 1692	Bandora and Tannara rivers, Bacaim	Outward voyage	-----
34	*Nossa Senhora da Ajuda e Sao Francisco de Borja or Lusitania*	1693	Mandovi bar, Goa near Aguada	Outward voyage	-----
35	*Anglesea*	23 Jul 1738	Wallwan ao norte de Goa, India	De Reino Unido para Madrasta, India via Bombaim	The ship was carrying Moeda
36	*Conceicao*	5 May 1767	Vengurla (Queimados Islands)	Goa to Bombay (Mumbai)	The ship lost at Queimados Islands, near Vengurla.
37	*Novo-Ceilao*	1807	Off Mangalore	--------	Wooden sailing ship; wrecked off Mangalore coast; part of the property saved from the wrecked ship.
38	*Chaldea*	22 Mar 1874	16 km(10 milhas) de Vengurla, India	De Bombaim, India para Calcutta, India	The ship was carrying Moeda

Table 1. Showing the Portuguese shipwrecks along the Indian coast

In the 17th century the naval war between the Portuguese and the Dutch off the port of Honnavar caused the loss of as many as 52 ships (Sewel, 1900). This must have happened because the Keladi ruler, Shivappa Nayaka, drove the Portuguese from Honnavar with the assistance of the Dutch army (Saraswati, 1925).

2

Coastal Geology and Geomorphology of Goa and Equipment and Methodology

Goa is part of the western coastal region of India. The physical characteristics of the state are similar to its neighbouring regions on the west coast. The most important geological feature is the extensive laterization because of its position in a moist tropical climate, subject to pronounced seasonal changes. Laterite caps are widespread over the terrain, mountains, plateau and plains. Goa occupies a 105 km long by 40 km wide strip bordered by the Arabian Sea and the Western Ghats. Several isolated hills rise in the coastal tract between the ghats and the sea. The larger part of Goa is occupied by the Dharwar Super group of Archaean Proterozoic age. The Pre-Cambrian rocks and the Deccan traps are covered by about 30 to 40 m of laterite on the coast which thins towards the interior. The presence of laterites is an important component of the geology of Goa. The older formations are covered by recent and sub-recent sediments. The rocks have been folded and intruded by granite gneiss, hornblende granite and porphyritic granite (Gokul et al., 1985). In addition to these rocks, a narrow strip of Deccan trap occurs at the north-eastern corner, while the presence of quartzite, probably of the Proterozoic age, occupied on the north-west region along the coast (Kamat, 1990). The presence of laterite beds forming abrasion platforms in the intertidal zone and water depths between 27 and 35 m below present sea level along the estuaries can be attributed to the quaternary rise of sea level giving rise to drowned valleys. Similar drowned valleys have been reported elsewhere along the west coast (Vaidyanadhan, 1968). Marine abrasion platforms are found at the base of the cliffs and headlands. It also appears that some lower level platforms are submerged.

The 105 km long coastline is marked by sandy beaches at times interrupted by rocky promontories or headlands. The coastal tract extends from the River Terekhol (Tiracol) in the north and Kalinadi Creek in the south. The coast consists of a combination of beaches, creeks, rocky shores and headlands which protrude into the sea. Sandy beaches separate the rocky headlands, cliffs, river mouths and wide estuaries which have shaped the coastline. The major rivers namely, the Chapora, Mandovi, Zuari, Sal, Tiracol, and Talpona originate in the Sahyadri ranges and join the Arabian Sea. The lower reaches of all these rivers are tidal. Hence during high tide the low lying adjacent areas submerged. Some of the beaches are backed by several rows of sand dunes. The headlands consist of basaltic and metamorphic formations capped by laterite and intruded by dykes and dotted with forts and lighthouses on their summits, namely at Aguada, Reis Magos, Chapora fort, Cabo and Cabo de Rama used for hinterland and coastal navigation (Gomes, 2004). The coastal region comprises a complex system of wetlands and lowlands, tidal marshes, cultivated paddy fields and intertidal beaches, all intersected by canals, inland lakes, bays, lagoons, creeks and other features regulated by (normal) tides. The rivers Chapora, Mandovi, Sal and Zuari define the drainage pattern of Goa (Iyer and Wagle, 1987; and Iyer et al., 1989.).

Seabed topography off Goa

The coastal and marine geology and hydrology are important environmental parameters, which are an essential part of maritime archaeological and shipwreck studies. The coastal belt comprising river basins, sandy beaches, estuaries, offshore bars, reefs, shoals, lagoons and islands including the geomorphology of the Arabian Sea. The coastal tract can be broadly classified into beaches, sea cliffs, promontories, estuaries, sand dunes; hard rock wave-cut platforms and sandy patches. The promontories of Bardez and Sashti created the estuaries of the Mandovi and Zuari rivers which provide a natural harbour and two anchorages. The coastline has bays and headlands which are intersected by the large estuaries of the rivers Mandovi and Zuari and at intervals is interspersed with minor estuaries. The headlands have played a significant role as a landmark for coastal and offshore navigation (Gune, 1979). Strongholds were built on the headlands of Aguada, Cabo and Cabo de Rama during the Portuguese colonisation.

The seabed topography suggests that there is likely to be a number of shipwrecks off Goa because the topography of the ocean floor plays a key role in causing shipwreck. Similarly, topography, sediment-distribution, currents and wave characteristics in a particular location affect the degree of preservation of shipwrecks. Marmagao harbour is sheltered even during the south-west monsoon and serves as an all weather port, whereas Aguada was once a fuel and water supply point but is not used today. The coastline also consists of a narrow strip of sandy plain. Hydrographic charts show that the continental shelf is relatively wide. The coastal belt is characterised by a number of geomorphological features related to marine erosional and depositional processes. The seabed consists of silty clay at water depths of less than 50 m and sandy silt from a depth of 50 to 100 m. Longshore and tidal currents, accompanied by strong winds blowing from the Arabian Sea, have played an important role in shaping the coastline (Wagle, 1982). Evidences of sea level changes in the form of marine platforms and beach rocks in the coastal belt show that at one time the sea encroached on the plains. St George's Reef, Grande and Pikene islands appear to be detached portions of coastal headlands, now isolated from the receding coast as a result of erosion or rise in sea-level. Even the present Marmagao headlands were once an island. They are now connected to the mainland by a peninsula on which stands the town of Vasco-da-Gama (Wagle, 1982).

The Sunchi Reef is located between the Marmagao and Cabo headlands and oriented in a north-south direction across the entrance to Marmagao Bay which separates it from the Arabian Sea (Fig. 4). Bathymetric contours indicate that the area has a gradual slope towards the west. The River Zuari with its wide opening to the sea near the Amee Shoals in

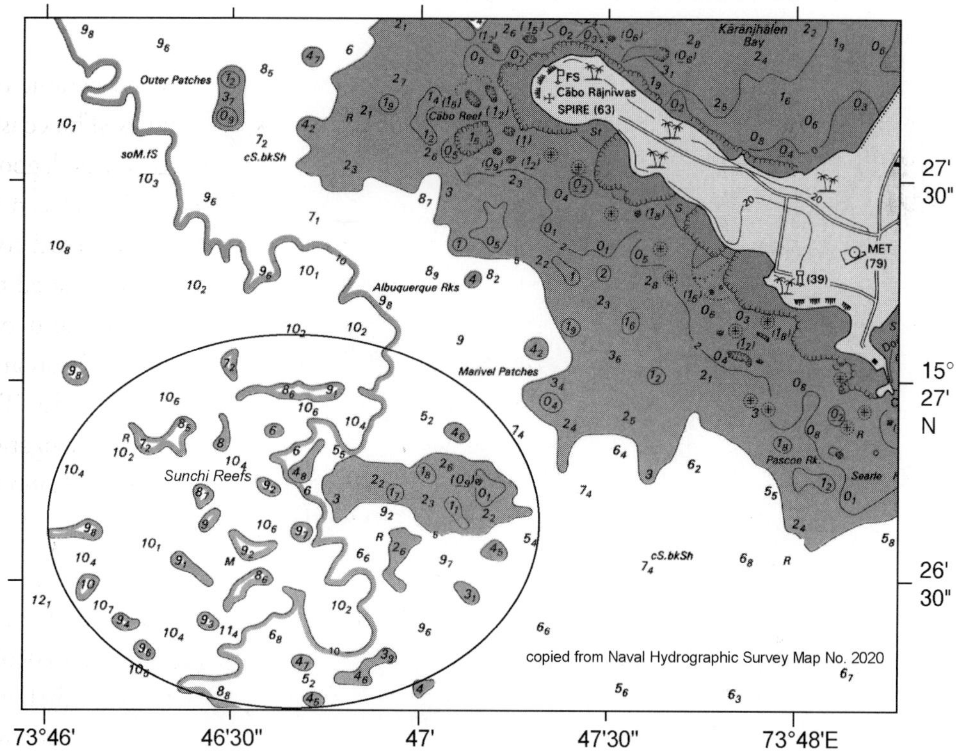

Fig. 4: Location of Sunchi Reef

the west is the main supplier of sediment to beaches in both the nearshore and offshore regions. The Sunchi Reef consists of laterite shoals more than 5 m in height. These rocks extend onshore (Ambre, 1991) and are hazardous for navigation. The reef near the shipwreck site consists of an uneven laterite shoal at a water depth of 3 to 9 m. The currents, along with wave action in the reef region, helped to deposit a thin layer of silt on the remains of the wreck. The water depth in the centre of the shoal is less and gradually increases in a north-west direction. The maximum speed of ebb currents is about 3 knots, whereas during a spring tide it is about 1.8 knots. Coarse sand and sediment extend towards the sea from the north-west end of the bay as a strong current prevents the settlement of fine material. As the Sunchi Reef area is very shallow, the current action in the working area is strong and severely hampers work. The seabed here is covered with shelly sand, lithified sediment and a rocky bed thinly covered with sediment (Wagle et al., 1988). St George's Reef is located on the eastern side of Grande [Grandi] Island and to the south of Marmagao port (Fig. 5). St George's Reef submerges during the highest spring tide and has an underwater extension on its southern side. The reef does not extend below water on the western side towards Grande Island. However, during the lowest neap tide the entire surface of the reef can be seen clearly. The seabed around the reef comprises coarse sand, varieties of shell and fine silt

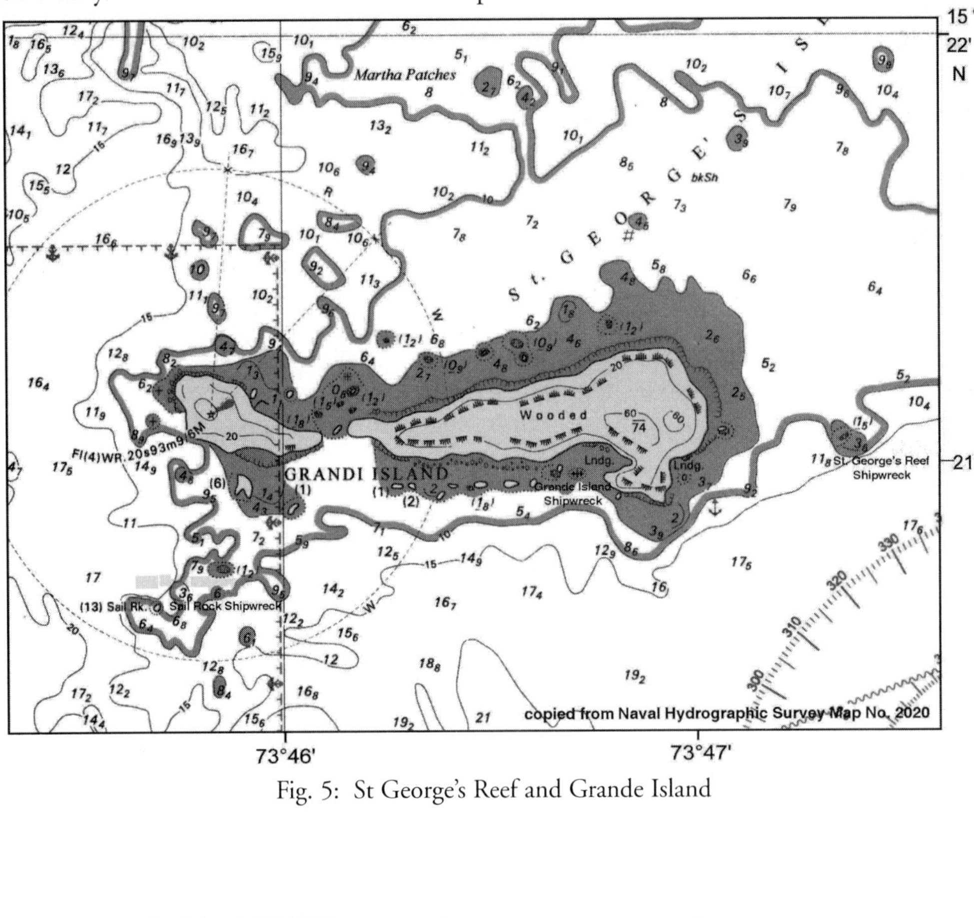

Fig. 5: St George's Reef and Grande Island

and towards the north coral, gorgonian, sponge growth can be seen. The reef is a navigational hazard. The depth at the work site varies on account of the gentle gradient on the southern and south-eastern side of the reef. Sometimes vigorous currents and winds churn the sediment and reduce visibility. The Amee Shoals lie in front of Marmagao Bay and on the southern side of Sunchi Reef, oriented in a north-south direction. The Amee Shoals separate the bay from the Arabian Sea. Bathymetric contours of the Amee Shoals indicate that the area has a gradual slope towards the west. The River Zuari joins the sea near them to the west. The offshore geology and other parameters of the shoals and the surrounding region have been studied in detail (Veerayya et al., 1981). An overgrowth of soft corals and gorgonian are noticed on the reef. The nature of the seabed is silty and a thin layer of suspended material can be seen on the laterite. Coarse sand, sediment mixed clay and sediment extend towards the sea from the north-western end of the bay as a strong current prevents the settlement of fine material. As the nearshore area lies in a high energy zone, the absence of a deposit of clay and silt causes faster biofouling and deterioration of artefacts. The ebb currents are 3 knots and tidal currents vary between 1.5 and 1.8 knots during spring tides (Pathak et al., 1991). Crevices are present on the laterite. Some shipwrecks in the Amee Shoals area are marked on Hydrographic Chart No: 2020. Maximum water depth in this area varies from 2 to 9 m.

Equipment and Methodology

Boat (Survey Vessel)

The seabed topography of the nearshore zone of Goa is uneven, a laterite outcrop is present and the water depth is less than 4 m. This prevents the deployment of a coastal research vessel such as '*Sagar Shukti*' or regular fishing trawlers to undertake diving and underwater exploration. A fibre hulled canoe is found quite suitable and is hired for exploration purposes in Goa waters. The canoe which was hired was 8 x 2.5 m with a work space on the afterdeck suitable for keeping diving gear and also space forward to keep cameras and other equipment. In addition, a rubber inflatable 'Gemini' was also deployed. But in the case of investigation of the shipwreck off St George's Reef a standard sized wooden hulled boat (fishing trawler) was chartered. The vessel had ample working space on the afterdeck which was suitable for diving operations and other equipment, while the cabin space was occupied by an underwater video monitor, still cameras, metal detectors and other equipment.

Diving operations and Diving gear

Scuba diving is most suitable for the exploration and documentation of shipwrecks. Scuba diving was initiated at wreck sites and was gradually extended to surrounding areas. Each dive team covered a large area (about 5 to 10 m radius) in free circular and swimline search patterns. Sometimes diving continued for more than an hour. Water depths varied from site to site. After the discovery of wreck remains, cleaning, drawing and photography of any objects study of the environment of the site, critical for underwater documentation purposes, was undertaken. At places hand fanning was also carried out to expose any antiquities buried in sediment. It is important to have accurate positions of shipwrecks and other remains, so the Global Position Fixing System (GPS) (Fig. 6a) is used.

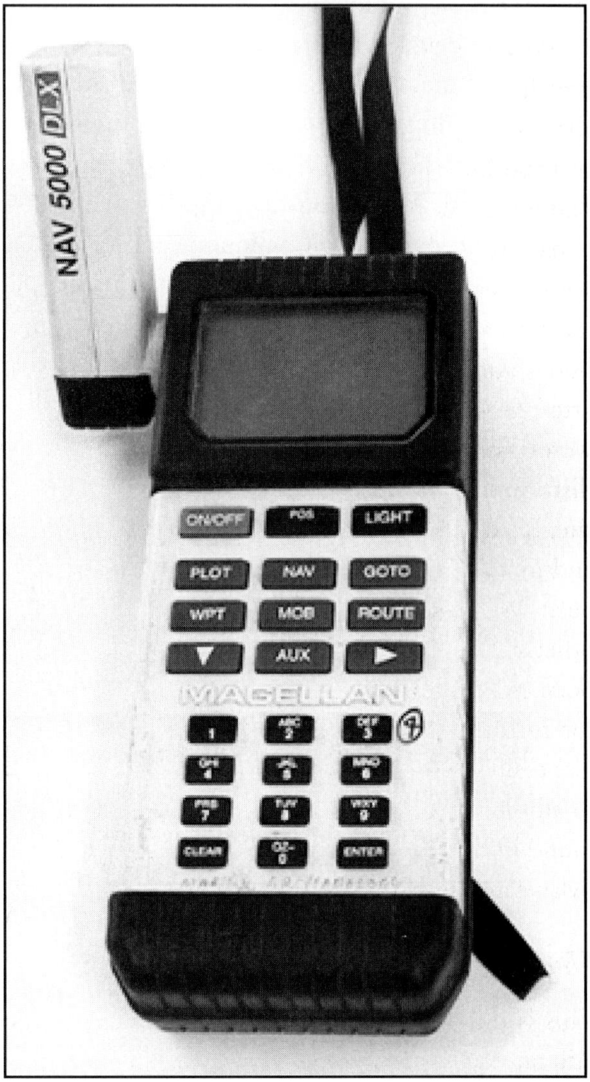

Fig. 6a: Handheld GPS

Scuba diving gear (Self Contained Underwater Breathing Apparatus) consists of one or more portable cylinders to carry compressed air, a demand valve which assures the supply of air for the diver at a comfortable pressure (the pressure a diver requires is the same as that of the surrounding water), a buoyancy compensator device [BCD] to hold the tank on the diver's back and an air pressure gauge to indicate the tank pressure. The depth gauge shows depths at all levels and also records the maximum depth the diver reached during a dive. The mask provides a clear vision under water. The fins aid the forward movement of the diver and the weight belt gives the diver negative buoyancy in order to undertake a specified job. Without weights, a diver would not be able to descend. In India a wet suit is found more suitable than a dry one because of the tropical climate. A wet suit is designed to fit the body snugly but allows water to penetrate at the neck, wrists and ankles, forming a thin film over the entire body. This film quickly reaches body temperature and thus acts as an insulator against cold. It also prevents further cold water from entering. The system enables the diver to dive freely without hindrance. Scuba is simple, yet remarkable, in that it delivers air on demand at a pressure slightly above the ambient pressure. The Scuba system is extremely useful for underwater survey especially when operating an underwater scooter, underwater TV and still cameras, drawing, using a communication system or a hand-held metal detector.

Underwater Exploration

Underwater exploration is generally based on known archaeological and historical data which requires visual inspection of a site and its surroundings without disturbing any of the artefacts. Visual seabed survey provides full information on the nature of the seabed, distribution of the shipwreck and state of wreck remains. Swimline and circular search patterns are found suitable for shipwreck survey; the same method is followed and attention is paid to understanding the distribution and state of archaeological objects lying on the seabed. The entire area was explored in order to discover the extent of the wreck site. If the location and positions of earlier finds were known, then those were relocated and marker buoys are tied to them. Subsequently, further survey was carried out, and wherever new artefacts were discovered, marker buoys were attached. All the artefacts found under water were interconnected with polypropylene rope and their direction, distance, area of distribution and degree of the same are obtained in order to understand distribution on the site. In addition, natural features like rock outcrops, seabed topography, flora, fauna, depth and movement of sediments and current direction in the region were noted. As ships have been wrecked on the reef and shoals, more emphasis was given only to diving, because other surveys might not yield better results than diving. Smaller objects or artefacts such as potsherds and metal objects belonging to the wreck, which could provide important clues for further research, were retrieved for study and analysis after proper documentation. Generally, underwater investigation begins at a known point on the wreck site then extends to adjoining areas. It is important to have the exact position of wrecks and other remains found in the sea; hence the co-ordinates are taken by Satellite-based Global Position Fixing (Magellan model NAV 5000 DLX) System.

Underwater Photography

This is one of the most useful recording techniques for underwater archaeological sites. Still photography is essential for recording finds relatively more quickly than other methods of documentation. A photographic record allows subsequent detailed study of the site, objects and general setting of the area of a wreck site. Still cameras (photography) were deployed for documenting shipwreck and other archaeological remains. An underwater Nikon V Camera with 15 mm wide angle lens was used for still photography on Sunchi Reef. For artificial lighting a Strobe Gun of 1000 candlelight was used. A scale of 100, 50 or 25 cm length depending on the size of the object was placed near the artefacts and antiquities being photographed. The composition required for every photograph of each object is different. Konica and Kodak 100, 200 and 400 ASA colour films or 100 and 400 Ilford Pan B/W films were used. Similarly, the Nikon RS camera was also used for underwater still photography. The Nikon RS camera consists of two main units, namely the camera which holds the lenses, and a 35mm film roll. It is a fully automatic camera; otherwise it can be used manually. The other part of the camera consists of different lenses, such as 50 mm, 35 to 70 mm and zoom. The camera can take pictures from 0.4 inch, and in case of good visibility up to 15 feet. In good visibility 35 mm lenses can cover about a 15 foot area. The strobe light of the camera provides good pictures. The strobe charges within 5 to 8 seconds. Though the weight of the camera is 8 kg, due to its positive buoyancy its weight reduces to half under water (Fig. 6b-c).

Fig. 6b: Nikon V underwater camera

Fig. 6c: Nikon RS underwater camera

Fig. 6c: Nikon RS underwater Camera with flash

Colour watch Underwater Inspection system and Sony Handycam

The Colour watch Underwater Inspection system consists of an underwater camera, colour monitor, keyboard, recorder and an umbilical cable. The unit was kept on board and one could see the picture of underwater objects on the monitor, which is attached to the camera by an umbilical cable (Fig. 6d). The system converts images captured under water into a

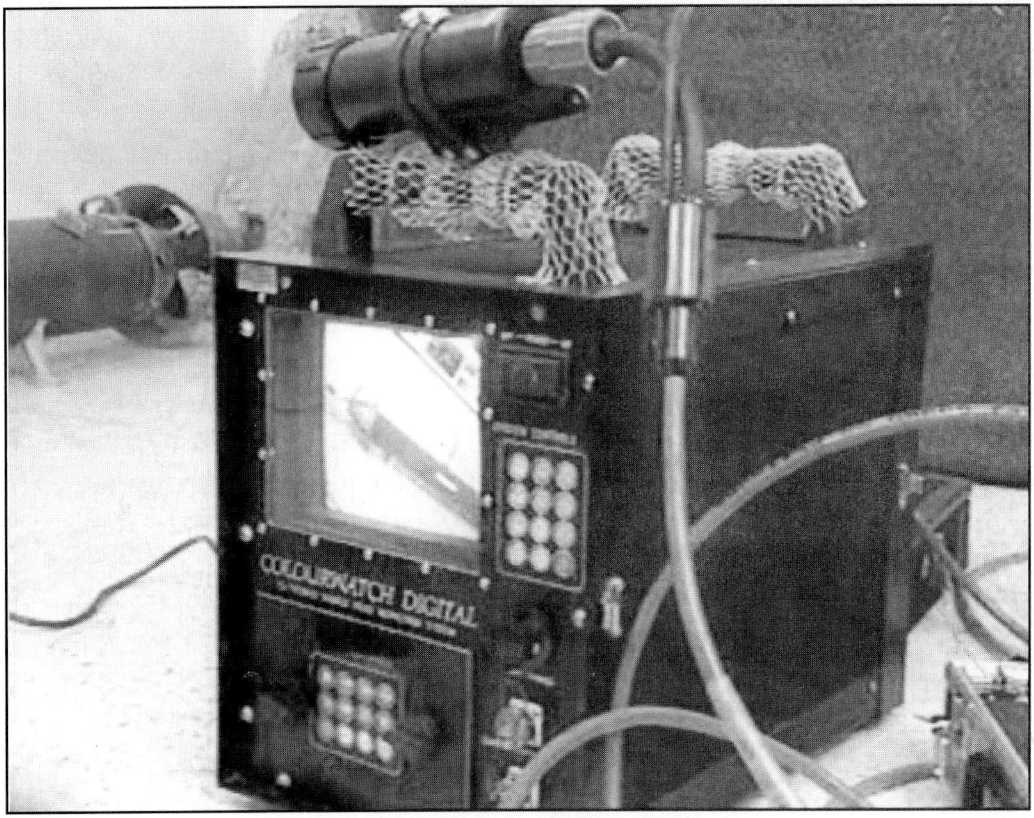

Fig. 6d: Colour watch video camera

digital image through a decoder, and records on mini DVC (digital video cassette). The monitor has a light control, which can be illuminated under water in three modes, i.e. low, medium and high, depending on the requirement. The other options available with the system are audio record DV and VHS output. Even conversations with the diver can be recorded through the underwater communication system. The camera unit is fitted with a water resistant chamber. The unit can be deployed at a depth of up to 100 m. All focus functions are available on the monitor. The diver merely takes the camera into the water. There are various lenses available ranging from 15 to 72 mm. In good visibility the camera can cover an area of 7 to 10 sq ft and in poor visibility only up to 2 sq ft. Required information such as date, depth, name of the site and diver, location and other details can be typed using the keyboard writer and the same information can be recorded on the DVC if required. Another tool used for recording maritime archaeological surveys and documenting the artefacts, structures and other finds is videography by using a standard movie Sony Hi 8 camera. This system was also used to record finds where it was found difficult to use still photography because of poor visibility or easy access in the working site. The normal movie Sony Hi 8 camera is used for underwater survey. This camera was used in Sony housing and is deployed under water for recording purposes. The advantage of this system is that a diver can move freely anywhere under water and record at will. The camera can be deployed at a depth of as much as 40 m. The quality of recording depends upon visibility, sea conditions and the nature of the site.

Underwater Drawing

The basic method followed for documenting underwater discoveries is underwater drawing or sketching. Generally, underwater drawing is carried out during good visibility so that the object can be seen clearly from all angles. A drawing is prepared on Garware polyester tracing film, which is fixed on a suitable sized aluminium drawing board with a graph sheet placed below the tracing film (Fig. 6e). Ordinary H or HB pencils are used for underwater writing. The artefacts are drawn *in situ* along with accurate measurements, bearing and description. Sometimes the date, time, water depth, object, location and direction are noted on the drawing, of help while finalising the drawings. The same drawing is finalised

Fig. 6e: Underwater drawing

for publication. Besides a drawing board, an underwater slate made of half a PVC pipe or an acrylic board are used to draw underwater finds. They can be tied on the hand with a Velcro (hook and loop fasteners) (Fig. 6f). The advantage of the slate is that it remains to

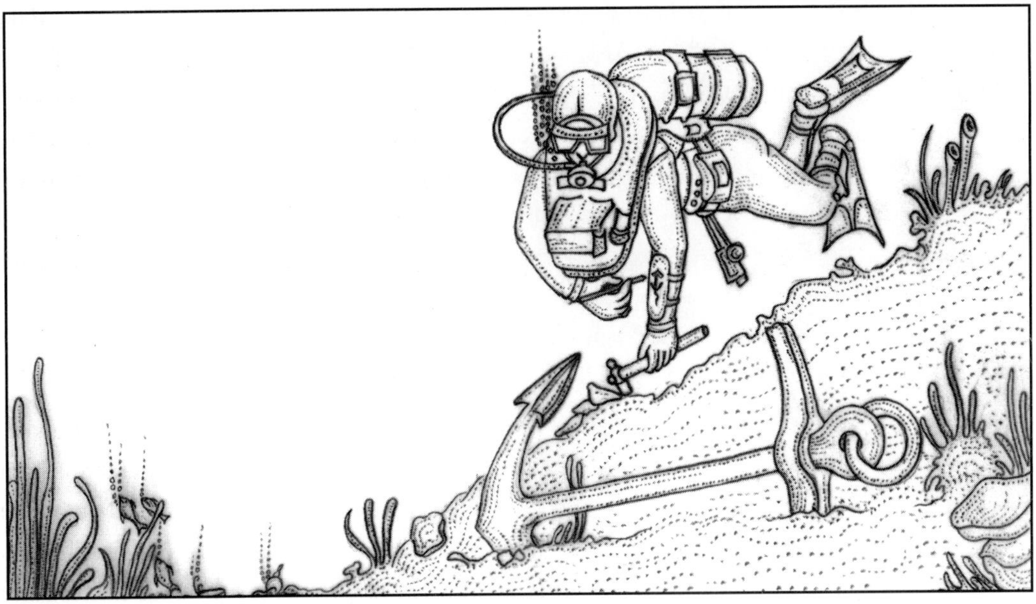

Fig. 6f: Underwater slate

hand and writing on it is very easy under water. Importantly it does not hinder the diver's movements.

Underwater Excavations (airlifting)

Comparable with land excavations, underwater excavations are carried out either by airlift or water-dredge, which are used to remove sediment by suction to expose buried objects. A compressor on board or ashore builds up air pressure in a container to about 90 lbs per square inch. The compressed air is carried through a rubber tube to the seabed where it enters a nozzle attached to the excavator's PVC pipe of 10 to 15 cm diameter. The length is adjustable in order to reach the seabed. It is suspended from the surface, the lower end being made of flexible material so that the operator can manoeuvre the nozzle over a wide area. As the compressed air shoots up the pipe it creates a partial vacuum in the nozzle, thus creating a suction which draws into the pipe water, loose material, small artefacts and potsherds from the seabed. The presence of compressed air in the nozzle makes it very light in weight, so the operator sitting astride it can control the apparatus properly (Fig. 6g). The power of suction depends upon the amount of pressure injected and the water depth at the

Fig. 6g: Airlifting

work face. While operating the airlift, the diver holds the lower end of the PVC pipe and takes care that small, fragile artefacts are not damaged but carefully collected for study. Sometimes after draining the water the spoil lifted on board is searched for archaeological remains and then returned to the sea.

Underwater Metal Detectors (boat-towed and hand-held)

Metal detectors are classified as remote-sensing tools. Unlike magnetometers they can detect both ferrous and non-ferrous metals. Metal detectors are generally used during pre-disturbance survey of a site to locate metal objects which are not visible due to poor visibility or burial in sediment. For exploring a large area a boat-towed metal detector is found highly efficient.

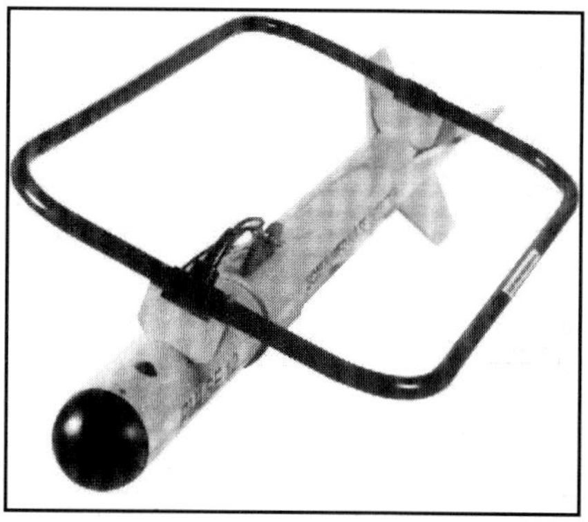

Fig. 6h: Boat-towed metal detector

The Pulse 12x underwater boat-towed pulse induction metal detector consists of a 0.6 m by 0.90 m coil, a towable fish, a pre-amp, a 100 m cable and a control box. For underwater use, the coil is mounted on the fish and is towed at any depth down to 100 m. The fish swims through the water at a depth determined by the speed of the boat and tows the coil very level and directly below the boat (Fig. 6h). Pulse detectors operate by transmitting a stream of high energy magnetic pulses (one hundred per second), from the coil. After each pulse is transmitted, the unit then 'listens', using the coil as the receiving antenna. When a transmitted pulse hits a metal object, a magnetic field is induced in it. This causes eddy currents to flow in the metal, which in turn generates a second magnetic field. This field is picked up by the coil, amplified and then displayed by the meter and heard in the speaker as well as the earphone. The Gemini (rubber boat) navigates visually on a grid of approximately 3 to 4 m. However, the rubber boat has no direct positioning system. It is intended to cruise at a very low towing speed (0.5 to 1 km), so that when a ping is recorded by the metal detector, a marker buoy can be dropped immediately to pinpoint the source. Subsequent verification can be carried out by hand-held metal detector and visual inspection by maritime archaeologists. It will detect small targets though not at great distances. Its major advantage is that it is very sensitive and does not generally detect minerals (extremely high concentrations of iron ore may give a modest reading). This insensitivity to minerals is particularly useful for salt water divers who may have to contend with heavy mineral deposits on the ocean floor producing large false readings on conventional detectors. The Pulse 12x with its very large coil, has a Sample Pulse Delay Knob which enables the operator to tune out most minerals. The sensitivity of the Pulse 12x is switch selectable and is dependent upon whether it is operating in fresh or salt water. When operating in salt water, a second switch sample pulse-delay is used to adjust the salt water impact on the large pulse 12x coil. The sample pulse delay setting does affect the detection range on smaller targets, but has little impact on larger ones. The detection range depends on whether the targets are buried under mud, coral, sand or rocks. The detection envelop for the pulse 12x is much larger than the coil width and length. Pulse 8x is the hand-held metal detector which is carried by archaeologists to search for buried metal objects in the seabed (Fig. 6i).

Fig. 6i: Handheld metal detector

It is a watertight sealed chamber housing a coil powered by a rechargeable 14-volt battery. The charged battery runs for four hours at a stretch. It has the capacity to detect any kind of metal within a range of 1.5 to 2 m. This can be used up to a water depth of 50 m. The range increases if the object is larger. A hand-held metal detector is very handy and useful in shallow water surveys.

Global Positioning System

The Global Positioning System (GPS) is a recent introduction in maritime archaeology. Before its invention the sextant and Miniranger were used to locate work sites. The sextant is a navigational instrument used for measuring solar and stellar heights, whereas in Miniranger the electronic position fixing system works in the 'C' band on a line-of-sight principle. When the Miniranger is in operation the mobile station transmits microwave signals at 5570 MHz. Magellan model NAV 5000 DLX GPS has been used in survey to obtain the position of explored areas, both shipwreck sites and port remains. This system collects data from satellites on five channels to compute position, elevation, velocity and navigational data. Once a satellite signal has been located, it takes about 12 to 30 seconds to collect complete data on a location. The accuracy of the NAV 5000 DLX is from 25 to 100 m. The other advantage of GPS is that if the geo-coordinates of the site are already available then the coordinates are entered and guided to use the location and give the distances. In order to obtain the accurate position of the site regular readings are taken for some time at intervals and the mean value is computed. The co-ordinates of explored areas and shipwreck findings were taken by Satellite based Global Position Fixing (Magellan model NAV 5000 DLX) system.

Location of shipwreck sites

Indian epics, *Puranas*, legends, folklores and traditions refer to shipwrecks. Archival documents and marine records which are housed in various archives and libraries in India and abroad mention shipwrecks in Indian waters. Similarly, Indian ships wrecked in foreign waters are also mentioned. These records provide information such as the name of ship, destination, captain, port of call, cargo, causes of wreck and other details. Most of the records remain vague about the exact location of a wreck. In the case of Goa, information on shipwrecks was collected from marine records housed in Goa State Archives, Panaji, and India House, Lisbon. Published sources, scholars local fishermen and divers were also consulted. In the course of examining marine records housed at Goa State Archives information on several shipwrecks was collected and records mentioning ships wrecked in Goa and adjoining waters because of hidden rocks, shoals, bars and severe storms (Gudigar

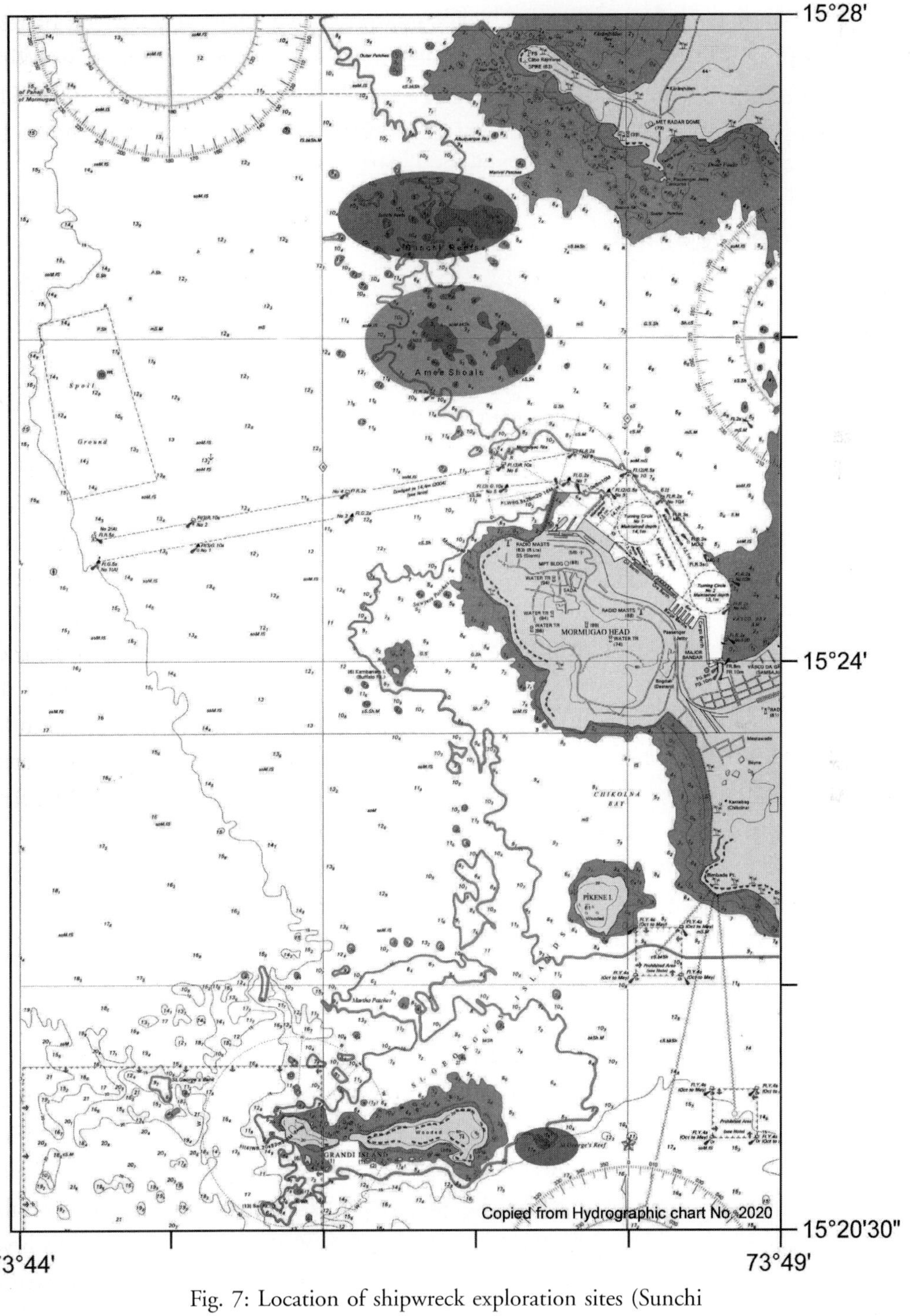

Fig. 7: Location of shipwreck exploration sites (Sunchi Reef, St George's Reef and Amee Shoals) off Goa waters.

et al., 1995-96; and Thakkar, 1991a). The Portuguese records mention that between 1497 and 1612, 860 ships voyaged from Portugal to India but barely 426 ships returned (Mathew, 1988). Shipwrecks are also marked on hydrographic Charts. It was deduced that exploration of the shipwreck sites marked on hydrographic charts might add information on the shipwreck archaeology of Goa. Hydrographic Chart No. 2020 shows shipwrecks in the Amee shoals area. The Amee Shoals are approximately 1.5 km south of Sunchi Reef which lies behind Cabo and Dona Paula (Fig. 7). The whole area is filled with shallow patches of rocks and water depths vary. Preliminary investigation of shipwrecks on Sunchi Reef was carried out in 1989. However continuous exploration has been undertaken from 1997 onwards. At Sunchi Reef, shipwreck was re-located and attention was paid to understanding the distribution and state of objects lying on the seabed. Natural features like rock outcrops, seabed topography, depth and movement of sediments and current direction in the area were noted.

St George's Reef lies on the eastern side of Grande Island 1 km away from Grande Island; the reef completely submerges during high tide. The water depth in the surrounding area is more than 10 m. The shipwreck off St George's Reef lies in 15 m of water but is recorded neither in marine records nor on hydrographic charts. Another steel-hulled wreck found off Grande Island, although the wreck can be seen above water. In the recent past, exploration of the Amee Shoals brought to light a steam engine shipwreck and the shipwreck scattered over a large area. The remains are documented and boiler bricks and flanges have been retrieved from the site for study and analysis. In addition to the survey of shipwrecks on Sunchi Reef, St George's Reef and the Amee Shoals, underwater survey has been carried out off Aguada Bay and Old Goa. In Aguada Bay, exploration was carried out opposite to the circular bastions of the fort to locate either shipwrecks or iron anchors and was restricted to a specified area. At Old Goa exploration was confined to a limited area in the River Mandovi between the passengers' jetty and Pentagon Engineers boat repairing yard.

Limitations

The west coast of India experiences high wave activity during the Southwest monsoon. However, it remains relatively calm during the rest of the year. The current action off Sunchi Reef, Amee shoals, Aguada bay and St George's Reef during the ebb tide is quite strong because of tidal variations and shallow depth. These high energy zones cause problems for archaeologists undertaking maritime archaeological work. The flow of current here is up to 0.6 m/s, the wave height during the Southwest monsoon rises to 5.1 m and the tidal range is 2.3 m (SanilKumar et al., 2006). Similar problems were also encountered at Old Goa while working in the River Mandovi.

Visibility

The major rivers of Goa, namely the Mandovi and Zuari discharge huge quantities of sediment into the sea. This sediment derives from mining activities inland. Aerial photographs of the offshore region show the suspended sediment load. Goa provides a synoptic view of discrete water masses which can be differentiated by sediment load. In addition to the coastal turbid water zones, high turbidity water can be seen entering the Arabian Sea. The visibility in Sunchi Reef, St George's Reef and the Amee Shoals region becomes poor because of the presence of fine sediment on the seabed and on laterite boulders. In shallow water during low tide and rough sea the bottom sediment often gets churned up, resulting in decreased visibility. Underwater documentation, such as photography, videography and drawing, become extremely difficult under such conditions. This problem was also faced during visual survey at Aguada and Old Goa where suspended materials do not allow sunlight to reach the seabed. In the nearshore zone of Goa visibility can be assumed on the basis of the colour of the water. Whenever the wind blows in a west to east direction, the swells increase, wave-height exceeds 1.5 m, and reduced visibility makes any task difficult. This occurs almost every season. Winds are stronger during the summer monsoon and the strength of the winds decreases after it ends. The sea remains calm during the morning hours as the day advances the wind gets up. Visibility improves only if the wind blows from a south to north direction. During low tide the visibility reduces to zero and diving becomes impracticable. However, underwater exploration at high tide yields good results.

Sedimentation

The rivers Zuari and Mandovi are the major sources of supply of sediment to the beaches and offshore regions. In addition to other rivers, the two major rivers carry sediments from mainland mining activities. The sediment supply from these rivers increases during the Southwest monsoon between June and September. The sediment of Sunchi Reef, the Amee shoals and Aguada Bay area appears to have been carried by wave action and rivers discharge sediment directly at these sites. There are bottom eddy currents here which constantly churn up sediment. The fine sediment is poorly sorted and consists predominantly of silt and clay. The Sunchi Reef, Amee Shoals and Aguada Bay region is full of rocky and sediment deposit can be noticed on shipwreck remains. On St Georges Reef the deposition and movement of sediment is greater than on Sunchi Reef.

Marine life

Environmental factors such as bottom composition, currents, sea animals, plant life and visibility affect diving and underwater investigations. Marine animals such as jellyfish, stingrays, eel, venomous fish, hydroids and sea wasps are problematic since they can inflict injury to divers, fatal at times. Light orange-coloured jellyfish which are abundant badly hamper diving operations. Sometimes exploration has to be suspended due to their presence. Some jellyfish species cause an adverse reaction in humans. Contact with tentacles can trigger millions of nematocysts which pierce the skin and inject venom. Jellyfish frequently come in contact with divers and cause painful boils on the skin.

3

Exploration and Excavation of Shipwreck off Sunchi Reef

In archaeology, the study of artefacts, such as coins, medallion, pottery, seals, sculptures and metal objects, is important in order to identify their age, provenance, and the context in which they are found. Similarly, in nautical archaeology the study of shipwrecks and shipwreck finds is important, because in-depth study may provide information on what caused the shipwreck, the origin and history of the ship and other details. For instance, its interaction with people, precise construction and history before it was wrecked can often be obtained. In general terms 'shipwreck' denotes 'material remains of a ship' which has either sunk or beached. Shipwreck study is known as nautical archaeology in order to cover fresh as well as sea waters. 'Nautical' is derived from the Latin *nauta* - 'sailor', It thus has a broader compass than 'marine' and 'maritime' archaeology, both of which are derived from the Latin *mare* - 'sea'. Nautical archaeology covers men and the whole environment in which they operated, in particular the design and construction of their water transport, whether merchant vessels, warships or fishing boats. It includes derelicts and vessels subsequently buried under dry deposits. Shipwrecks occur due to many causes, such as submerged rocks, reefs, sand bars and shoals, while violent weather or naval warfare may cause a vessel to founder. Whatever might be the cause, a wreck is an instance of a physical event. The United Nations Organisation (UNO) has estimated that there are more than three million shipwrecks lying on the ocean floor. Maritime or nautical archaeologists are motivated to locate shipwrecks which can preserve information about long distance maritime trade, seafaring, warfare, life on board and contemporary shipbuilding methods. The job of the maritime archaeologist is to create an accurate picture of the past from available clues either from shipwreck remains or from artefacts found under water. Shipwrecks are important because they enhance knowledge of maritime history, trade and cultural contacts of a particular period from the material remains - hence 'shipwrecks' are considered to be 'time capsules'. Shipwrecks are essentially maritime casualties, but constitute part and parcel of the cultural heritage of a country. As a ship can be wrecked in foreign waters its cultural context will relate to other countries than that where it ended up. Shipwrecks are contributors to world cultural heritage, not that of any one country.

In the recent past, thousands of wrecks have been explored and excavated both in marine and fresh water around the world. After sinking, deterioration of cargo and component of ship begins due to natural forces and marine organisms. The decay process continues for a period then diminishes gradually to the point where it is disturbed by natural forces, such as erosion, or human intervention by dredging or diving. The process of decay is then reactivated. In the 20th century, pioneering divers surveyed and excavated the submerged ports, trade centres and shipwrecks of countries other than their own in order to reconstruct past trade and cultural contact with other countries. Nautical archaeology is thus truly international from its inception.

A number of developed and developing countries subsequently made significant initiatives in maritime archaeological research. India has made significant progress in shipwreck archaeological studies and its potential has been demonstrated by shipwreck exploration in Lakshadweep (Rao et al., 1995-96; and Gaur et al., 1998; and Sila Tripati and Gudigar, 2001), Poompuhar (Gaur et al., 1997; and Sila Tripati et al., 2003c) and Goa waters. The exploration of shipwrecks off Goa became a priority because of the wealth of documentary information. Details on shipwrecks off the Goa and Konkan coast has been collected (Gudigar et al., 1992; and Gudigar et al., 1995-96). Shipwreck data along the rest of the Indian coast has also been compiled (Fig. 8). Taking into account available information from archival records, hydrographic charts and local divers the NIO carries out surveys to locate and excavate shipwrecks. After the inception of maritime archaeology in India the exploration of shipwrecks began in Goa waters in 1986-87. Preliminary surveys took place off Grande Island and subsequently off Sunchi Reef. Responding to the importance of the shipwrecks in Indian waters, the Department of Ocean Development (DOD) (currently known as Ministry of Earth Sciences), Government of India, extended financial support to survey and excavate wrecks on the west coast of India. As a part of the DOD funded project, exploration was carried out off St George's Reef, Sunchi Reef and in Vijaydurg waters of Maharashtra. The St George's Reef survey brought to light building construction materials belonging to the Basel Mission Company carried on a century old

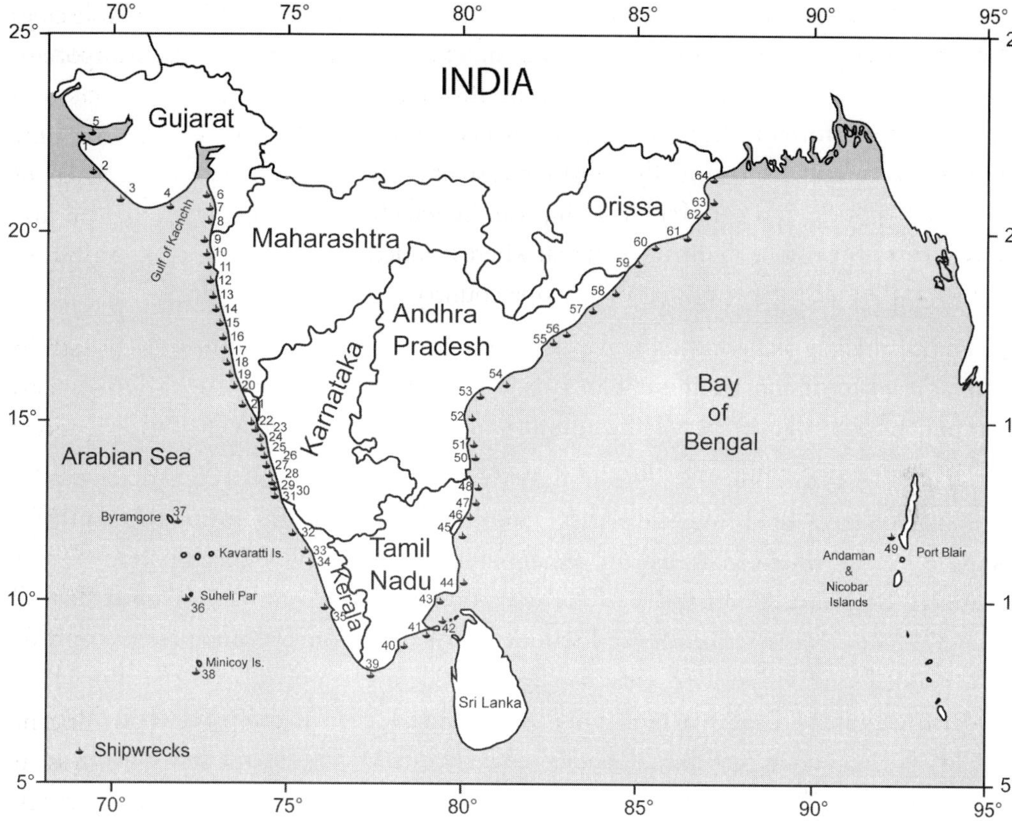

1. Dwarka
2. Miyani
3. Porbandar
4. Diu
5. Mandovi
6. Surat
7. Daman
8. Umargaon
9. Kolaba
10. Elephanta Island
11. Alibagh
12. Bankot
13. Jaigad
14. Ratnagiri
15. Rajapur
16. Jaitapur
17. Vijaydurg
18. Deogarh
19. Malwan
20. Vengurla
21. Goa
22. Karwar
23. Ankola
24. Honnavar
25. Murudeswar
26. Bhatkal
27. Kundapur
28. Malpe
29. Udipi
30. Mulki
31. Mangalore
32. Tellicherry
33. Calicut
34. Beypore
35. Cochin
36. Suheli Par
37. Byramgore
38. Minicoy
39. Kanyakumari
40. Tuticorin
41. Rameswaram
42. Gulf of Mannar
43. Nagapatnam
44. Cuddalore
45. Pondicherry
46. Mahabalipuram
47. Madras
48. Pulicut
49. Andaman & Nicobar Islands
50. Krishnapatnam
51. Nellore
52. Coringa
53. Masulipatnam
54. Narasapur
55. Visakhapatnam
56. Bimlipatam
57. Kalingapatnam
58. Sompeta
59. Sunapur
60. Gopalpur
61. Ganjam
62. Puri
63. Konark
64. Hukitula Bay

Fig. 8: Figure showing shipwrecks along the Indian coast

ship. The Sunchi Reef survey revealed iron guns, an iron anchor and dressed granite blocks. Subsequently, exploration was carried out off Sunchi Reef, the Amee Shoals and Baga.

Topography of Sunchi Reef

Sunchi Reef lies between Marmagao harbour and the promontories of the Cabo headland which extends in north-south direction and separates Marmagao Bay from the Arabian Sea. The sea floor of the area is characterised by a prominent laterite rocky bed with flat surfaces and sandy patches. Coarse sand and sediment extend seawards from the north-west end of the bay as strong current prevents settlement of fine material. At places, the height of the laterite shoal is more than 4 m. The ebb current in the Sunchi Reef area is 3 knots and 1.5 knots at springs. The entire region is hazardous for navigation.

The River Zuari flows into the head of Marmagao Bay through a constricted passage and a huge quantity of sediment discharges into the sea particularly during the Southwest monsoon. At this time the sea usually remains very rough and wave heights reach 4 m in the Sunchi Reef area. Comparison of bathymetric charts 1881 and 1976 shows that the Sunchi Reef area is not a sediment trap. Sediments collected from the sand patches comprise coarse grained sand mixed with various types of shell and fragments of oyster. Big laterite boulders are found with crevices of various sizes in which archaeological artefacts are trapped. Visibility in the Sunchi Reef area depends upon the tide, flow of wind and sea state.

Exploration of Shipwreck

The shipwreck off Sunchi Reef was discovered by local divers in 1975. The wreck is neither recorded in Portuguese records nor detailed on any charts (Fig. 9). On the basis of this information, a preliminary survey was initiated in 1988-89 to explore the wreck and document the remains. Subsequently, work continued between 1997 and 2006, only possible during fair weather and in good visibility. During ebb tide the speed of the current in the reef region is much stronger and it hinders under water survey and reduces visibility. On account of this, sometimes the duration of underwater exploration had to be curtailed according to the tide and visibility. Mostly work was carried out between November and March when conditions were most favourable. Survey and documentation was carried out by scuba. The wreck remains are scattered over a large area at a depth of 3 to 6 m. Some of the remains are overgrown with white and orange coloured gorgonians, green mussels and some places are covered with a fine layer of silt and coarse sand. During visual survey much attention was paid to the distribution of artefacts. In Certain places hand fanning was also carried out to expose artefacts buried in the sediment. Subsequently, underwater survey extended from the known area to the adjoining areas. The positions of finds were plotted

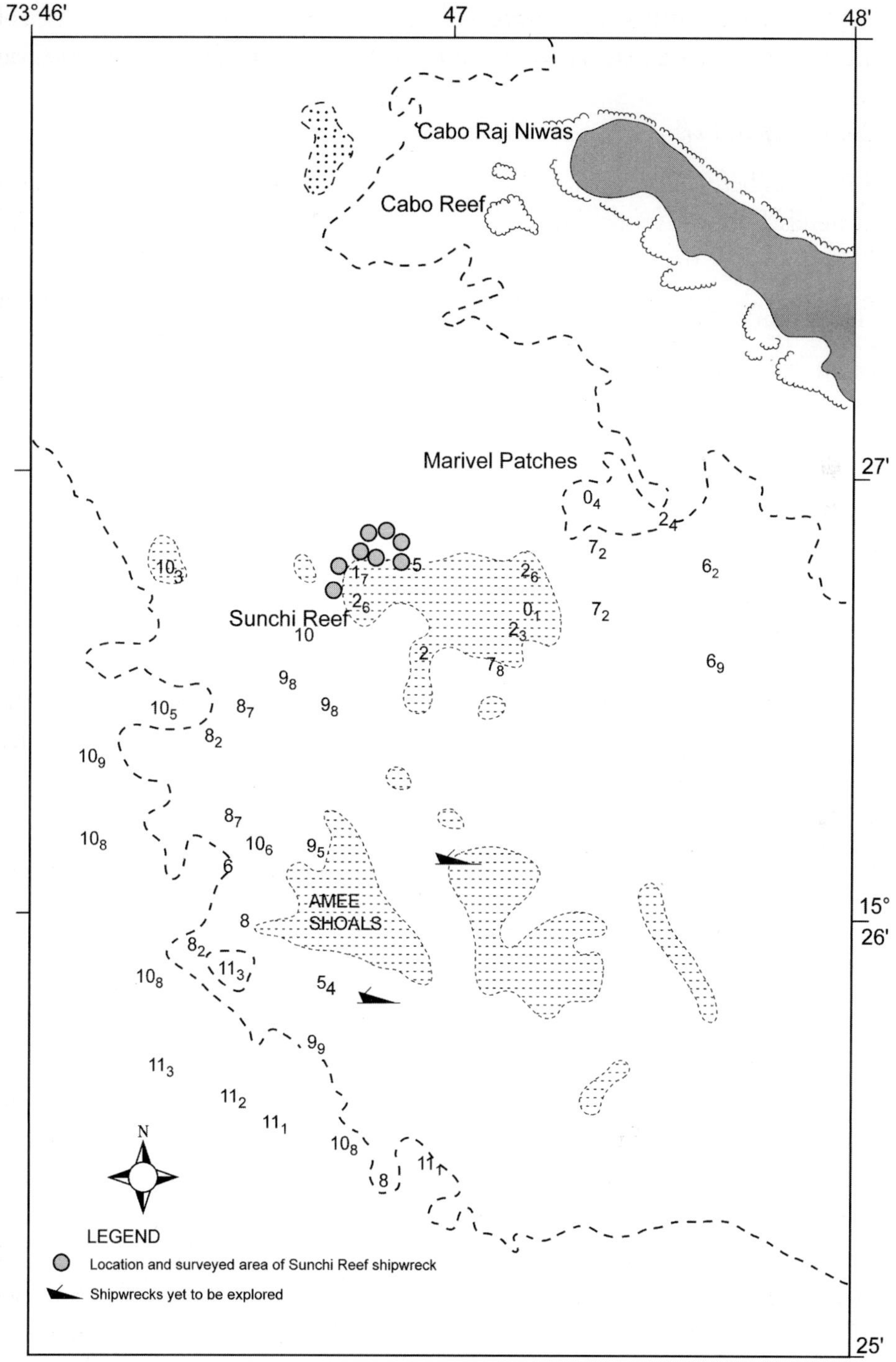

Fig. 9: Location of shipwreck off Sunchi Reef

using GPS and the smaller artefacts were recovered for study and analysis. More emphasis was placed on visual survey because the wreck remains are scattered over the reef. Due to the rocky seabed the use of an airlift was impossible there. Wherever new artefacts were found marker buoys were tied to them and positions obtained. Other features, such as outcrops of rock, the seabed topography and the movement of sediments and current direction were noted. Documentation and recording was the next stage, ensuring the exact position of underwater explored areas and location of any shipwrecks.

Shipwreck Finds

The survey of Sunchi Reef brought to light iron guns (Cannon) , an iron shot, a brass barrel of a handgun, an iron anchor, elephant tusks, hippopotamus teeth, sherds of storage jars, sherds of Chinese ceramics, bases of glass bottles, a metal hook, lead sheet, copper vessel, copper strip, bricks and dressed granite blocks. These finds were recovered after extensive and thorough search and hand fanning. The iron guns, iron anchor and granite blocks lie on and in crevices of the laterite and are covered in prolific marine growth.

Iron Anchor

The 'Admiralty Pattern long shanked' type of iron anchor from the ship was found to the north of the guns. This type of iron anchor was extensively used during the 16th-18th century on Portuguese naval and merchant ships. The discovery of an iron anchor suggests that the ship is unlikely to have been of Indian origin. Generally, ships of that period carried a number of iron anchors. However the Sunchi Reef shipwreck has yielded only one iron anchor. She may have dragged her other anchors before she ended up on the reef or her other anchors and her valuables may have been salvaged. The maximum surviving length of the anchor is 310 cm. The shank is clearly visible whereas the flukes lie between laterite boulders. The anchor does not have its wooden stock, which either rotted or was washed away. The swivel ring of the anchor, used to secure the cable, has completely eroded. The length of one fluke is 110 cm and the other fluke lies in a crevice in the laterite. Thick barnacles, mussels and other marine growths are attached to the anchor. Due to this it was not possible to determine if there was any coating on the anchor originally. Anchors of a similar size and type have been recovered off Aguada Bay and do have a preservative coat for protection from corrosion and marine organisms. Two comparable anchors are on display in the Archaeological Museum in Old Goa (Sila Tripati et al., 2003a).

Armaments

A number of naval battles were fought among European powers for supremacy in the Indian Ocean. Due to the frequency of confrontation all European vessels required cannon of various types together with other armaments for self-defence. To deter pirates and to resist attack by local rulers during this period a bare minimum of armament was also needed by cargo vessels. During the survey four cast iron cannons were discovered lying close to each other on the laterite boulders (Fig. 10). The approximate length of each gun is 200 cm. The muzzles of two of them point towards the seabed at an angle whereas the other two lie parallel to the seabed. Over time, the exterior surfaces of the cannon have become heavily encrusted with barnacles and mussels; hence monograms, symbols and

Fig. 10: Cannons lying on laterite boulders, Sunchi Reef

other identification marks are not visible, nor is it possible to measure their calibre (Sila Tripati et al., 2004). The trunnions are visible as are the cascabels, but no lifting eyes or rings in contrast to forge welded cannon which have very visible iron rings. Their vents are yet to be uncovered. It appears that they have two reinforcements and a cascabel and were once mounted on carriages and were used to fire at close range. Their range was probably between 1000 and 1200 yards and they are likely to be 12 or 18 pounders. They are identical and belong to the 17th or 18th centuries. No additional information will be available until they are lifted and conserved. Similar iron cannons are displayed in the Archaeological Museum (ASI), Old Goa and can be seen in forts around Goa.

As there are cannons on the site, naturally there are cannon balls. Cannon balls could be made of stone, cast or wrought iron. A cast iron shot has been retrieved from the western side of the cannon. It is slightly conical and the surface is rough (Fig. 11). Its diameter, specific gravity and weight are 954 mm, 477 cm and 1218 gm respectively. Using the radius of the shot (477 cm) it may be concluded that the original weight would be between 3300 and 3500 gm or 6 to 7 pounds (7. 25 lb). By means of its present specific gravity, the original weight of the shot is calculated to have been 2016 gm. It may have fitted a 4 pounder. The original density could be higher because corrosion reduces density. Neither identification nor stamp marks were noticed on the shot which has lost a lot of iron in the marine environment. As a result of having been immersed for so long water has been absorbed by the shot; this was removed subsequently by heating it in an oven. Dehydration cracks on the surface have resulted while the interior is completely corroded. Fungal growth as well as a black, sulphurous material, possibly gunpowder, was noted in the interior. There is no visible fuse hole and it may be a solid shot. Shots of similar kind are kept in the ASI Museum, Old Goa and belong to the Portuguese period. A small piece of iron was removed for Scanning Electronic Microscopy (SEM) (Fig. 12) and Energy Dispersive

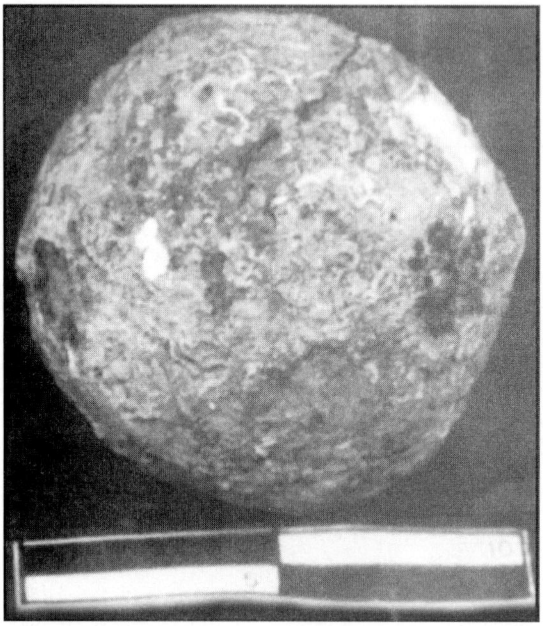

Fig. 11: Gun shot collected from shipwreck, Sunchi Reef

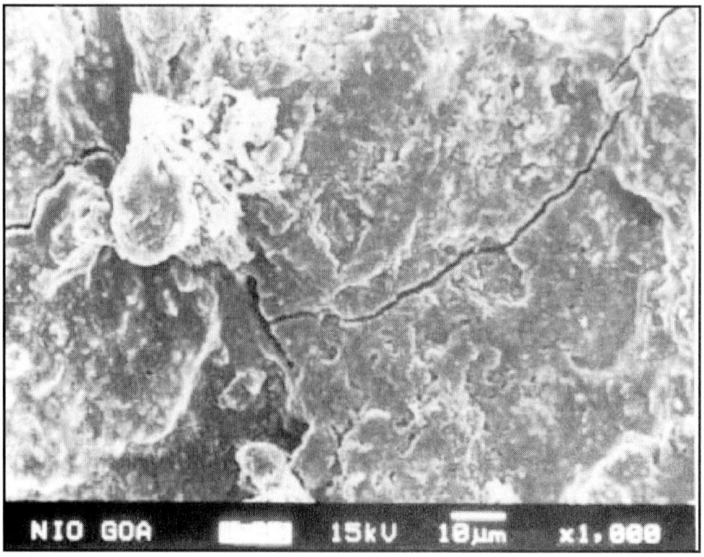

Fig. 12: SEM analysis of Gun shot: part of cast iron gun shot under low magnification, showing large cracks and its porous nature

X-ray analysis (EDS) to establish its chemical composition. The SEM examination shows the shot to be loosely bound, porous and cracked. The EDS analysis shows that the metal varied in iron content and other oxide percentages (Fig. 13; Table 2). The iron oxide (Fe_2O_3) content varies from 77 to 96 per cent iron with impurities of oxides of Si, Ca, K, Al, and Na; the variation in the oxide percentage could be attributable to changes in the sea floor environment or to the manufacturing process (Sila Tripati et al., 2001).

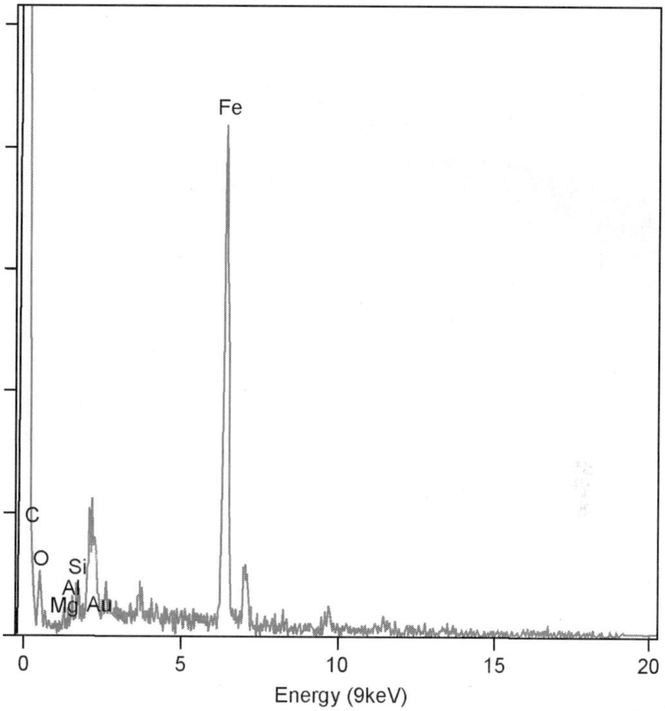

Fig. 13: Energy Dispersive X-ray analysis (EDS) graph showing chemical composition of gun shot (AU = gold coating of the sample)

Compound	Percentage (1)	Percentage (2)	Percentage (3)	Percentage (4)
Fe_2O_3	95.98	93.57	92.00	77.53
SiO_2	—	2.91	4.81	5.76
CaO	—	2.20	—	—
Na_2O_3	—	—	—	9.83
Al_2O_3	—	—	—	5.98

Table 2. Chemical composition of the cannon ball found on Sunchi Reef

Further exploration of the Sunchi Reef area revealed the brass barrel of a handgun, which was trapped and half buried below the dressed granite blocks. The front of the gun was buried in the sediment while a little portion of backside was visible. Its length and weight is 56 cm and 2200 gm respectively. The bore of the muzzle is 40 mm. It has a muzzle ring about 50 mm from the mouth. The barrel is thinner on the front than the back and little of the back is eroded. It has a screw thread at the back for attaching the firing

Fig. 14: Brass barrel of a handgun collected from shipwreck, Sunchi Reef

mechanism. The barrel has two notches for fixing screws to the wooden part provided below the handgun. Out of the two notches, one on the front side is intact and the other on the back is broken (Fig. 14). The gun and other parts attached to the barrel are missing. Marine growth is present on the surface and no marks are visible. No other parts of the gun were found nearby. Similar barrels of handguns have been found in Portuguese wrecks off Portugal (Fig. 15). In order to understand whether the handgun was made of brass or other copper alloy, a sample was taken from the barrel and analyzed using Energy Dispersive Spectrum analysis (EDS) equipment with a Scanning Electron Microscope (SEM) (Fig. 16). The analysis confirmed that the barrel was made of copper and zinc with minor varying amounts of Fe, Pb and tiny quantities of chromium (Table 3). The presence of a greater percentage of copper and zinc indicates that the barrel was made of brass. Whilst copper and zinc dominated the composition, even in this small sample, the Fe, Pb and Cr content varied considerably, indicating that in the process of manufacture this alloy was imperfect and homogeneity of the casting was not achieved. It is possible that the Pb derived from the gunpowder used.

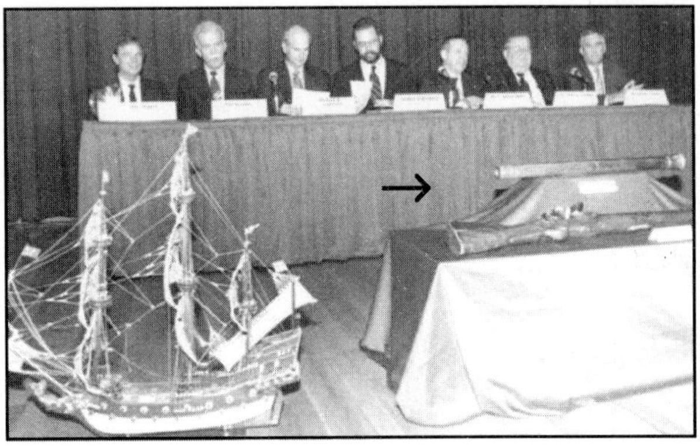

Fig. 15: Brass barrel of a handgun recovered from Portugal waters

Oxide	spot 1	spot 2	spot 3	spot 4	spot 5
Cr_2O_3	01.08	n. d.	n. d.	n. d.	n. d.
Fe_2O_3	20.41	07.33	02.35	0.74	0.90
CuO	46.56	52.72	52.99	55.55	55.55
ZnO	31.95	39.95	40.12	39.97	39.09
PbO	n. d.	n. d.	04.54	03.75	04.47
Total	100.00	100.00	100.00	100.00	100.00

n.d: not detected

Table 3. Chemical composition of the barrel of the handgun found on Sunchi Reef

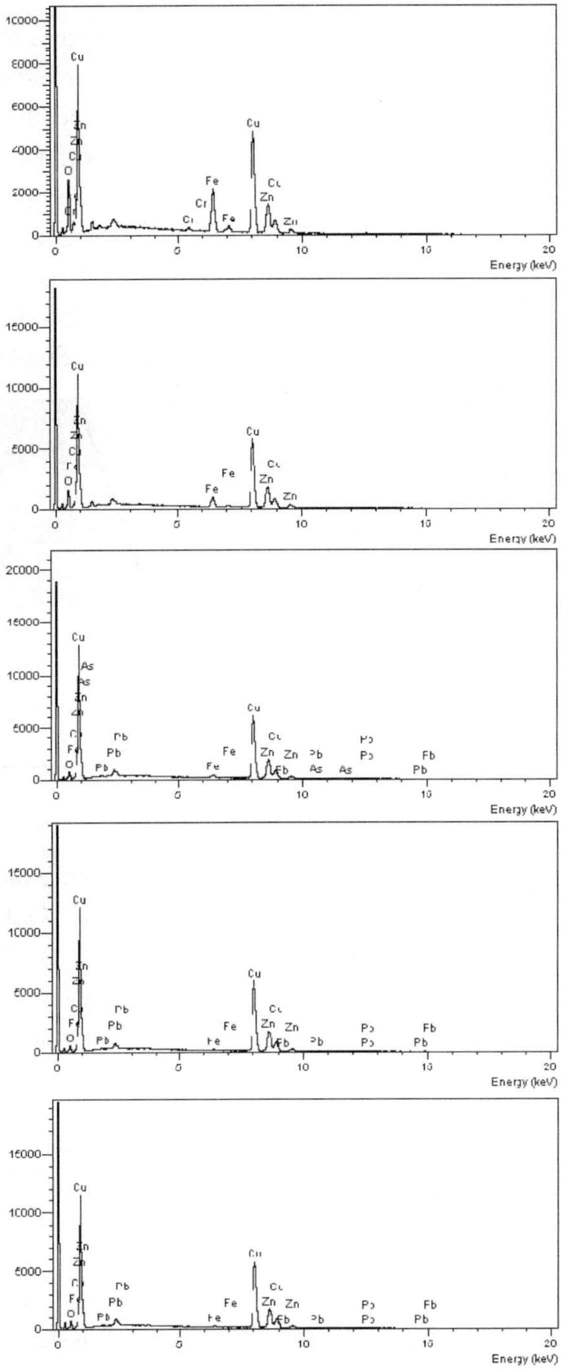

Fig. 16: Energy Dispersive X-ray analysis (EDS) graphs showing chemical composition of the barrel of the handgun

Caulking

Caulking was one of the most vital techniques and performed with the utmost care on ships, particularly those deployed on overseas trade ventures and naval activities. In the past the deck of a sailing ship was caulked with oakum, a tarred fibre made from old rope and cordage. It thus consisted mainly of jute or hemp fibres impregnated with pine tar, an amber coloured organic pitch derived from the roots of pine trees, not the mineral tar used in paving streets nowadays. Traditional shipwrights in India used to caulk vessels with oakum. The Italian traveller, Ludovico di Varthema (1470-1510), who visited India more than once and mentions that the shipbuilders of Kerala on the west coast of India did not use oakum between planks because they were joined so skilfully that water did not pass through (Jones, 1863; and Mookerji, 1912). In the early 16th century the Portuguese learned to protect a hull by applying additional planking both above and below the waterline, a technique which they learned after observing this treatment on Asian ships including Chinese junks. In addition to the use of oakum and provision of an additional layer of planking traditional shipwrights used lead sheets to cover the keel and hull. Sometimes the entire hull was covered with lead sheets in order to prevent its damage after striking against the submerged reefs and rocks. As extra protection, copper strips were nailed over the joints of copper or lead sheets and planking.

Lead pipe, sheet and Copper strip

The exploration of Sunchi Reef has produced a copper strip, three lead sheets and a pipe (length and diameter 375 mm and 360 mm respectively) from the wreck site (Fig. 17). The

Fig. 17: Lead pipe, sheets and copper strip collected from the Sunchi Reef shipwreck site

lead patches found were not uniform in size or shape. Of the sheets, two are 'L' shaped whilst the other is rectangular and plain. They do not have nail holes or marks. However, they do have tooth like projections on one side, which probably served as grooves for better fixing. Marine growth is present on all the sheets. They are very fragile, twisted and heavily worn. Lead sheet was used on board ship for a number of purposes: for cladding and inboard and outboard to plug leaks and effect emergency repairs after hitting rocks or reefs. Cladding if intact protected a vessel from infestation by teredo (the marine ship worm) and other organisms. During this period lead sheets were nailed on to the keel and copper strips were used to cover joints during construction. The single copper strip so far found is twisted and has a cut, probably where a nail or a bolt was fixed. The lead pipe found has some square holes, which are distorted and worn. Lead pipes were used as scuppers and as covers to protect hosepipes. Lead sheets and strip have been recovered from many shipwrecks including the *Nossa Senhora dos Mārtires,* wrecked in1606 off Lisbon (Castro, 2001; and Castro, 2005).

Cargo of the wrecked vessel

Elephant tusks and Hippopotamus teeth

Eight elephant tusks ranging from 32 to 65 cm in length have been recovered from the site. They were buried in coarse sand and covered with dead shells and only small portions were

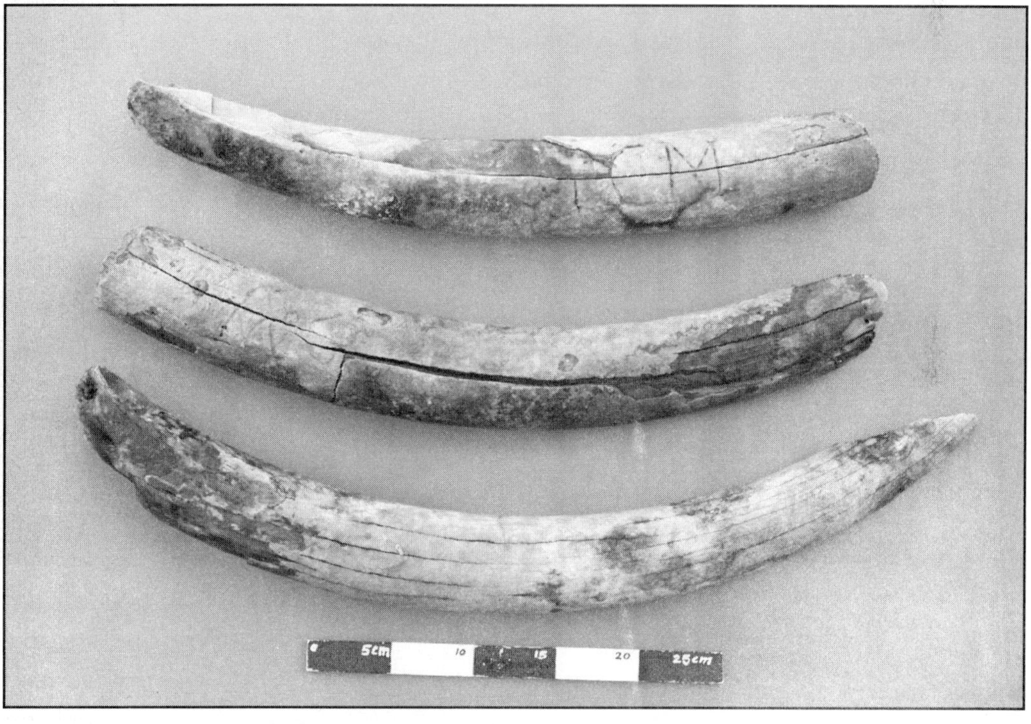

Fig. 18: Elephant tusk recovered from the Sunchi Reef shipwreck site

visible above the seabed. Two are inscribed. One tusk has three Roman letters, 'ICM', of which 'CM' is visible clearly (Fig. 18). They are possibly an acronym for a business organization or the initials of the owner. The second tusk has some geometrical designs. As a result of being under water for a long time the ivory has become highly degraded and

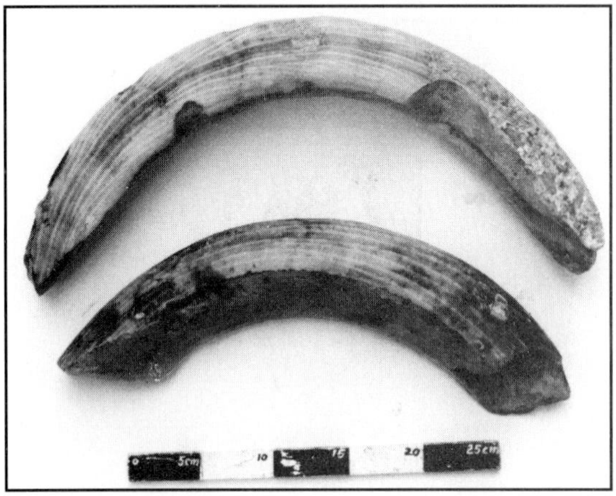

Fig. 19: Hippopotamus teeth retrieved from the Sunchi Reef shipwreck site

discoloured. The tusks cannot be allowed to dry out and are kept in fresh water. Local divers reported the recovery of a number of tusks of different sizes from the same wreck site (Sila Tripati et al., 2006a; and Sila Tripati et al., 2006b). Similarly, nine different sizes (535; 385; 205; 160 and 125 mm in length) of hippopotamus teeth (Fig. 19), partly buried in the seabed, were recovered from the area adjoining the guns. All are canines and curved. Incisor teeth, which are straight, were not found. Each canine has a straight and very smooth natural depression formed by continuous abrasion of the lower and upper jaw. Enamel is present on the outer surface, and hard dentine (ivory) on the inner side. Originally the teeth were white but have discoloured under water and have marine growth on them. In order to stabilise the tusks and teeth they were kept in marine water for a few days and then it was gradually replaced with fresh water. After desalination was completed they were stored in fresh water. A comprehensive report appears in Chapter 9 (151-162).

Chinese Ceramics

Chinese ceramic sherds consisting of bases, rims and a lid of medium fabric were found. Blue on white decoration is present inside and out. A base and another sherd have green decoration on white on the inside. A number of other bases were collected during exploration. Five bases have blue on white designs, two are plain and the other base has a green on white design inside. Chinese characters are present on two basal sherds; on one the character reads 'lucky jade' but its period is not specific. On the other there are illegible Chinese characters (Fig. 20), (personal communication: Roxanna Brown). The other sherds are painted both inside and out. Base sherds belong to jars, plates and bowls were collected.

Fig. 20: Stamped Chinese jar bases collected from the shipwreck off Sunchi Reef

Fig. 21: Chinese ceramic lid recovered from the Sunchi Reef shipwreck

Some bases have a raised ring and others do not. A small broken lid has blue on white decoration on the outside with a plain interior. The knob of the lid has floral designs (Fig. 21). Marine growth is present on all sherds. The majority of Chinese ceramics collected are broken and not a single intact jar, bowl or plate has been recovered. The reason is likely to be the impact when the ship struck the rock. On the basis of the Chinese ceramics it appears that the assemblage includes pieces of more than one date and is not earlier than the 17th century. Further study will elucidate their design and origin. Similar types of Chinese ceramics have been found in Old Goa. The designs include lotus flowers, pomegranate, pine trees, hills and mountains. A study of motifs gives an insight into the social and cultural life and religious beliefs of contemporary Chinese society. Most of the decorative elements and symbols originate in Chinese folklore, history, traditions, mythology and religious beliefs. Almost always there are underlying themes or messages. Generalised decoration is rarely used. Excavations of *Nossa Senhora dos Mārtires* (1606), *Santo António de Tanna* (1697) and other shipwrecks have yielded similar Chinese ceramics.

Storage Jars

Fig. 22: Chinese character stamped on the shoulder of the storage jar from the wreck site, Sunchi Reef

Long before the introduction of metal containers for carrying liquids and materials on board, different forms of pottery were used for transporting both liquid and solid. This is evident from finds both under water and on inland excavations. In India, the first evidence of pots on board comes from the Ajanta paintings dating to the 6th century AD. Naturally the shapes and sizes of pots changed over time. The Sunchi Reef survey brought to light a number of 'Martaban' storage jar sherds, which were lying in the crevices of the laterite and buried in the coarse sand. They included rims, bases, lids and body sherds and belong to pots of different shape and size. The sherds are well fired and are made of a medium or thick fabric. Due to long immersion in sea water, paint and glaze disappeared on some sherds while others still have a dark brown glaze coat of slip which is flaking. However, they do not show spalling, cracking or other defects (Sila Tripati, 2004; and Sila Tripati et al., 2006). Recent excavation yielded a shoulder sherd from a brown-glaze storage jar stamped with Chinese characters (Fig. 22). Due to abrasion the characters are illegible, but the jar may date to the 17th century, probably manufactured in southern China. Different types of rims have been collected from the site. Some rims have horizontal or vertical loop handles and others have both types on the shoulders. There is also a high necked jar with vertical loop handles and linear designs around the neck (Fig. 23). This type of rim was not found in previous seasons off the Sunchi Reef. Its vertical loop handles are bigger and more prominent than the horizontal loop handles. Some of horizontal loop handles are smaller and prominent whereas others are broader and less prominent. In addition to these handles almost every jar usually has raised dots in a circle close to the neck and luting lines noticed on some jars sherds. Besides rims and bases, body sherds were recovered with a dark brown glaze, grey fabric, honey-coloured glaze and liner designs. From the shape and size of the rims it appears that these jars were made for different purposes or manufactured by different makers. On the basis of the bases found, rims and body sherds can be classified into two groups: (i) stoneware and (ii) low fired stoneware. Exploration has brought to light three types of rim, namely: (a) low fired stoneware with dark brown glazed horizontal prominent loop handles, without raised dots, without neck, a medium mouth and big-sized jars, (b) raised dots, small horizontal loop handles with a small mouth, high neck, and medium-sized jars, (c) raised dots, prominent vertical loop handles with dark brown glaze, a wider mouth, high neck, and medium-sized jars. Among the sherds collected the largest belong to stoneware jars with raised dots around neck and body. Ropes were tied to the loop handles during transport to provide greater balance, prevent breakage and spilling. The largest bases of storage jars have distinct string (pontil) marks (Fig. 24) made when they were removed from the potter's wheel. The recovery of differing sherds indicates that the jars were manufactured in different

Fig. 23: Storage jar rim retrieved from the shipwreck on Sunchi Reef

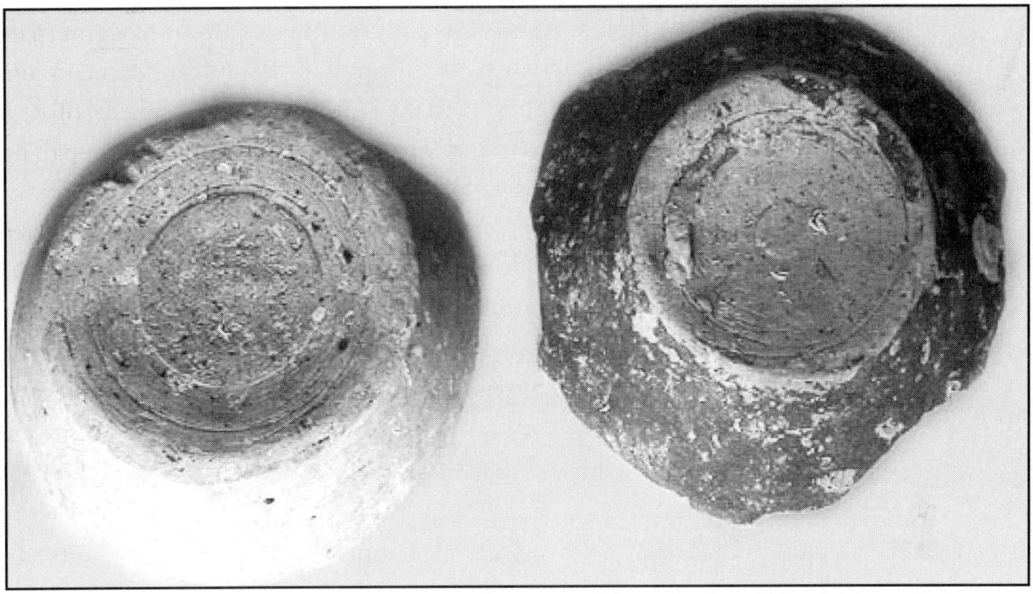

Fig. 24: Storage jar bases with pontil marks retrieved from the Sunchi Reef shipwreck site

parts of Burma, China and Thailand. The jar with Chinese characters shows its country of origin. The jars with dark brown glaze slip and horizontal and vertical loop handles may have been produced either in Burma or Thailand. In order to understand the mineralogy of these stone wares X-ray diffraction (XRD) analysis of three samples was carried out using a copper target on a (Rigaku Ultima IV) XRD machine. The power was run between 3^0 and 70^0 and the prominent peaks were identified following power diffraction data from JCPDS (Anon, 1981). Among the peaks obtained in the diffractometer only the peak of quartz could be identified (Fig. 25).

Copper Vessel

The exploration brought to light a fragile copper vessel (Fig. 26), which was trapped in the laterite. The vessel has been twisted and some portion is missing. It is made of single thin sheet without any rivets and does not have any designs or marks on it. The distinctive carination is present below the neck. It has a wider mouth and its height is less. The shape and size of the vessel indicate that it may not be suitable for kitchen and serving. Its particular use onboard is yet to be ascertained but it could have been used for various purposes.

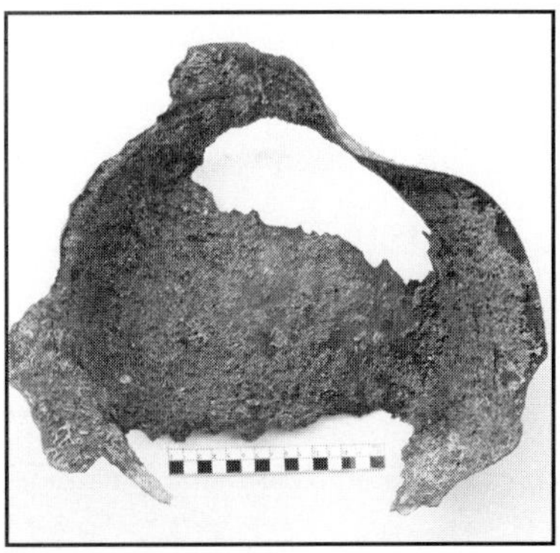

Fig. 26: Copper vessel recovered from Sunchi Reef

Peak Search

Sample	: 1	File	: 1.raw	Date	: November-13-09 12:07:48	Operator	: Administrator
Comment	:	Memo	:				
method	: 2nd differential	Typical width		: 0.000 deg.		Min. height	: 0.00 cps

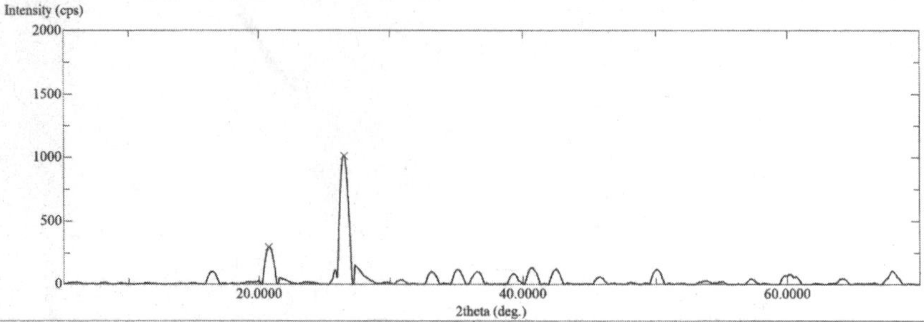

Peak no.	2theta	Flex Width	d-value	Intensity	I/Io	Peak no.	2theta	Flex Width	d-value	Intensity	I/Io
1	20.760	*****	4.2752	299	30						
2	26.580	*****	3.3508	1011	100						

Peak Search

Sample	: 2	File	: 2.raw	Date	: November-13-09 12:30:11	Operator	: Administrator
Comment	:	Memo	:				
method	: 2nd differential	Typical width		: 0.000 deg.		Min. height	: 0.00 cps

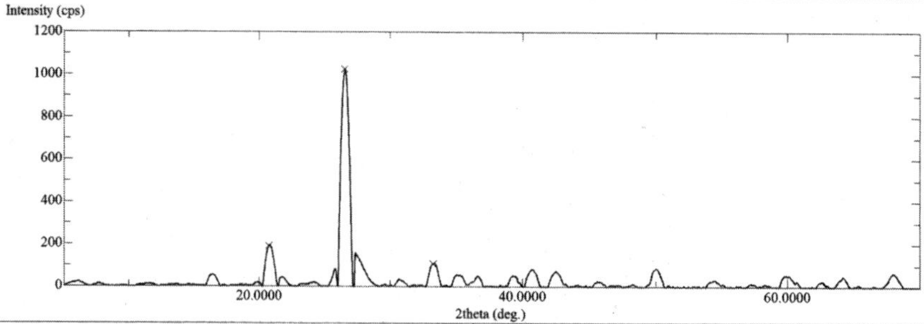

Peak no.	2theta	Flex Width	d-value	Intensity	I/Io	Peak no.	2theta	Flex Width	d-value	Intensity	I/Io
1	20.760	*****	4.2752	192	19						
2	26.580	*****	3.3508	1024	100						
3	33.240	*****	2.6931	106	11						

Peak Search

Sample	: 3	File	: 3.raw	Date	: November-13-09 12:52:31	Operator	: Administrator
Comment	:	Memo	:				
method	: 2nd differential	Typical width		: 0.000 deg.		Min. height	: 0.00 cps

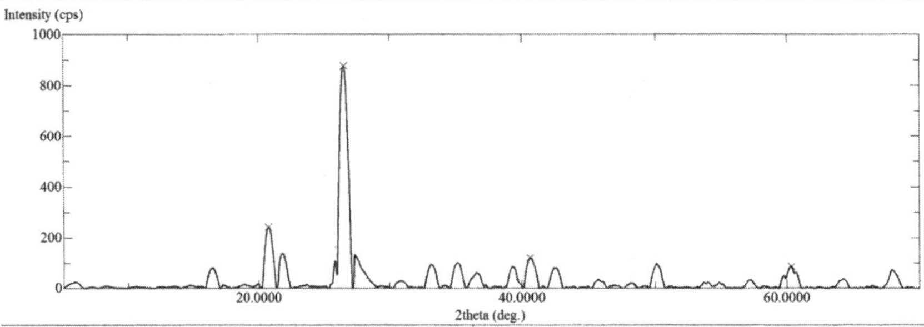

Peak no.	2theta	Flex Width	d-value	Intensity	I/Io	Peak no.	2theta	Flex Width	d-value	Intensity	I/Io
1	20.760	*****	4.2752	243	28						
2	26.580	*****	3.3508	876	100						
3	40.620	*****	2.2192	121	14						
4	60.340	*****	1.5327	87	10						

Fig. 25: XRD of stone ware sherds from Sunchi Reef

Door Knocker

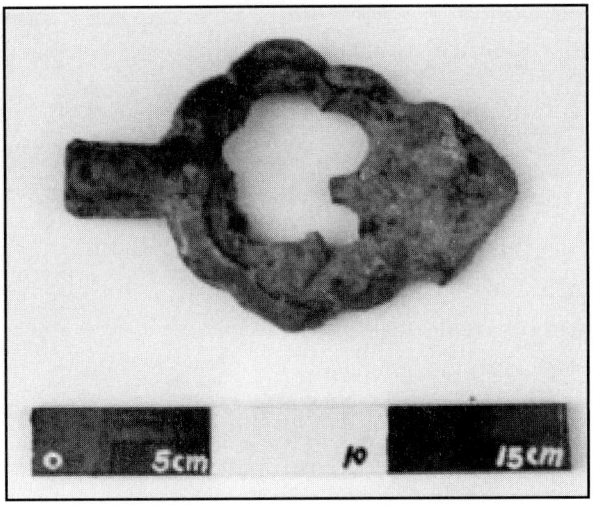

Fig. 27: Door knocker retrieved from the shipwreck on Sunchi Reef

Several types of door knockers have been recovered from shipwrecks. These door knockers not only serve the purpose of knocking the doors but also enchancing beauty. Even, some believe that door knockers protect against evil eyes to ships and houses, hence door knockers are common in ships. Varieties of door knockers have been reported from Portugal. A door knocker was found on the eastern side of the guns off Sunchi Reef (Fig. 27). It was buried in the sediment and a small portion was visible. Marine growth and sand accumulated on it. Its maximum length, width and weight are 12.5, 8 cm and 330.7 gm. It has no designs and signs on it. The knocker has a round hole at the lower side. The projected handles have holes meant for screwing on the door. The rear projected handle is bit broader and square than that of the front one. The front portion is round and bit smaller than rear one. It is an alloy of copper and zinc hence the bronze disease is present on its surface. A door knocker of bigger size with varied design has been found in Gibraltar, but it has been used in a private house of Laredo, Spain.

Bases of Glass Bottles

A number of broken glass bottle bases (Fig. 28) of dark green colour are recovered from the wreck site which were scattered over a large area. Among the collected bottle bases, sixteen have conical depression at the base, round in shape and their diameter is 28 cm, twenty are square bases which measures 8.5 x 8.5 cm and identical in shape. During the recent explorations a stamped square glass bottle base was found, which is illegible whereas on another base a geometrical design is present on the outer side. Maximum square bases have distinct pontil marks. Pontil marks are not present on round bases. In some of the round bases the push up depression are present. Marine growth is present on these pieces. The square bases are thinner in section comparatively round ones. Previous exploration has yielded a bottle with broken rim having round base stamped with 'A 0751' inside the depression, this could be either the makers' mark or the serial number of the concerned firm. On the basis of stamped mark this bottle appears to be of later period. Unbroken square base glass bottle so far has not been recovered from the site. It appears that olive oil, a traditional Portuguese product of export; spirit and probably special quality wine were imported in these glass bottles. Wine in large quantities was generally transported in wooden casks and bottled locally for retail use.

Fig. 28: Broken glass bottles bases (round and square) found in the shipwreck on Sunchi Reef

Bricks

Fig. 29: Stone and clay bricks recovered from the shipwreck on Sunchi Reef

Different types of clay, stone bricks and stone blocks were brought by the European ships to India as a part of cargo and ballast. However, Sunchi Reef shipwreck has yielded very little number of bricks which were neither part of cargo nor ballast. The underwater exploration yielded three well fired square clay and four rectangular stone bricks from the shipwreck site (Fig. 29). These are nicely trimmed from all sides. All the clay bricks are intact but profuse numbers of holes are present on them. They are well fired; rolled and marine growth can be seen on them. These bricks measure 15 x 10 x 3 cm, 11 x 11 x 4 cm and 10 x 10 x 3 cm respectively. Out of four stone bricks, two are intact whereas other two are broken. Out of the two intact, one is square and other is rectangular which measures 10 x 9.5 x 6 cm and 22 x 8.5x 5 cm respectively. These two broken bricks are flat, thin and different in sizes and raw material appears like slate. All these bricks belong to metamorphic group of rock. One of the possibilities of use on board the bricks must be for flooring, sometime these were used to cook the food on board for the crew and even used for covering the drainage.

Granite Stone Blocks

A large number of rectangular and square dressed granite blocks of different sizes were found lying at random on the seafloor of Sunchi Reef (Fig. 30). It appears that there could have been about 250 granite blocks, which were probably served both as cargo and ballast. Some blocks are more than 2 x 0.40 x 0.10 m in size. Maximum blocks have fallen in the crevices of the laterite boulders and concentrated in and around the guns. These blocks can be categorised under 4 groups based on their sizes such as 200 x 40 x 10 cm (232 kg), 190

Fig. 30: Neatly dressed granite blocks lying randomly on Sunchi Reef

x 35 x 15 cm (289 kg), 125 x 35 x 15 cm (190 kg), 75 x 35 x 12 cm (91 kg). The estimated weight of the above granite blocks have been calculated on the value of 1 cubic metre of granite is 2900 kg. The X-ray diffractometer (XRD) (Philips PW 1840) analysis of the rock sample shows the presence of quartz and feldspar as dominant minerals and corresponds to granite. Since ancient times stone blocks were carried on board as a part of cargo for various constructions and many shipwrecks have yielded stone blocks from shipwrecks.

Discussion

No documentary evidence has so far been found on Sunchi Reef shipwreck which could have provided some more information on the ship. On the basis of information from the local divers and fishermen, the wreck was explored and remains were documented. Some of the shipwrecks off Goa waters are shown in the Hydrographic Chart No: 2020 in the Amee shoals, on the southern side of Sunchi Reef and in the nearby areas of Grande Island. During the Portuguese regime, Old Goa was the capital, port and main trade centre of Goa. Ships coming to Old Goa used to cross the River Mandovi to reach Old Goa port. The sandbar near Aguada Bay at the confluence of the River Mandovi has caused grounding of a number of Portuguese ships. Sunchi Reef is not far away from the sand bar of Aguada

Bay. The captain of the ill-fated ship might have thought of sailing into the River Mandovi thorough Sunchi Reef instead of following the normal route. Though the causes of the shipwreck still remain a mystery however, the possible reasons for the wreck of Sunchi Reef could be the following (i) inadequate knowledge about the sea bottom topography of the region, (ii) the non-availability of navigational charts of the region showing the submerged rocks, shoals and reefs (iii) grounding of the vessel on the shallow submerged rocks because of human error in navigation and (iv) inclement weather. The location of the wreck clearly indicates that the ship struck the submerged rock and subsequently sank.

Topography, sediment distribution, currents and wave characteristics of a particular area play important role in the preservation of shipwrecks. The height of laterite cliff near the gun is more than 4 m and the current action during low tide in the region is very strong. The Sunchi Reef wreck is in Marmagao Bay and the region falls under high energy zone; shallow water topography covered with sand and shells and is not conducive to the preservation of the wreck. Any archaeological objects lying on the seabed will be exposed to marine environment thereby increasing the chances of corrosion and heavy encrustation. The lack of sand in Sunchi Reef area does not favour preservation of artefacts. Further, detailed studies in Sunchi Reef could provide an insight into how the ship wrecked and its destination. Except for a thin layer of silt and sand, not much deposit of sediment was noticed on the guns, granite blocks and iron anchor. However marine growth is more on the guns, anchor and granite blocks than that of potsherds and cannon ball. The dressed granite blocks brought on the ship were meant for some construction and also used as ballast.

As cannons are covered with marine growth it is not easy to decipher their types and make. Similarly, dating of cannons based on typology is also very difficult. All the guns were classified mainly by the weights of the cannon ball and the length of the gun. Guns range from 42, 32, 24, 18, 12, 6, 4 and 3 pounders. Average cannon ball weight was 24 lbs. These cannons normally kept on upper or middle decks, could be 4 pounders and fired from 1 to 1.5 km distance. Brass barrel hand guns have been recovered from the *Santo António de Tanna* 1697 and the *Grande Principessa di Toscana*, which are slightly different from brass barrel hand gun of Sunchi Reef. Some believe that the brass barrel hand gun of Sunchi Reef could be the barrel of a Musketoon (similar to a Blunderbuss) (personal communication: Robin Piercy) and early form of breach loading system. Blunderbuss was lighter than Musketoon, and at times these terms could be interchangeable. The lock of the gun would be either a flintlock but not a matchlock. These are often found on ships and were weapon of choice for boarding and close quarter fighting at sea. Its range could be 100 yards and sometimes these are supported on a rail of a ship, because of their heaviness. Barrels of hand guns have been recovered from the 17th century shipwrecks namely the *Ex*

Association and the *Eagle* which are datable to the 15th century AD (Morris, 1984); however, Brown (1984) has stated that the hand gun of the *Ex Association* could be of the 17th century but not 15th century AD.

The potsherds, glass bases and other remains are preserved as these are lying either below the laterite boulders or within the crevices and sometimes buried in the silt and sand. As the ship has been wrecked in rocky and shallow waters, the perishable material may have been decomposed and subsequently disintegrated, the light material such as glass bottles, tusks, ceramic, etc. were scattered over the region, some of them have been retrieved for study and analysis, whereas heavy materials namely the guns, granite blocks and iron anchor are lying *in situ* in their original location. The stone ware (also known as Martaban Jars) sherds, which have been collected from Sunchi Reef, were intended for carrying both solid and liquid items. All the rims of the jars have loop handles used for tying rope and the base of the jars has a distinct ring (pontil), which gives greater balance during transportation. Sherds of similar type and jars have been reported from the *Nossa Senhora dos Mártires* 1606 off Lisbon (Alves et al., 1998), *Santo António de Tanna* 1697 off Mombasa (Willoughby, 1991; Piercy, 1978 and 1981; Sassoon, 1981; and Moore, 1970), the VOC *Risdam* wreck off Malaysia (Green, 1986) and *Avondster* 1659 shipwreck off Galle harbour, Sri Lanka (Parthesius et al., 2003). For better preservation of eatables, sometimes wooden or clay lids were provided to close the mouth of the jars (Desroches, 1998). 'Martaban' has been named after the name of a city in southern Burma which was first centre of production and distribution of these storage jars. Subsequently, these jars were also produced in other places of Burma, China, Kampuchea, Thailand and Vietnam. Prior to the invention of modern type of containers, Martaban jars were used for carrying both liquid and solid material. These jars are fired under high temperature between 1100^0 and 1250^0 C (Van der Pijl-Ketel and Kist, 1982) hence they become vitrified, durable and strong like stone. Hence this ware is also commonly referred to as 'stone ware'. The height of these jars varies between 0.50 and 1 m. Further, they neither disintegrated nor decomposed even in marine conditions. Glazes often applied with an uneven consistency, frequently form into drips at the bases of these jars and were finished crudely. Usually, most of the jars have three, four, or five loops oriented horizontally with raised dots, sometimes vertical handles on their shoulders and these loops served as tie points to secure lids to the jars rims. Account of the 16th and 17th centuries and later accounts state that these containers were used for the storage and transportation of water, ginger, pepper, lemons and mangoes. These jars probably were also used as containers for grain, salted fish, wine, sugar, salt, tea, candied fruits, butter, oil, holy water and ceramics (Willoughby, 1991).

Comparatively, the findings of stone ware sherds are more in number than the Chinese ceramic sherds. The Chinese ceramics appear to have been used daily onboard because if the Chinese ceramics would have been part of cargo then a large quantity of ceramics would have been found *in situ* and adjoining area of the wreck. Sometimes the Chinese ceramics were also imported for trade. There are still lots of research yet to be done on the ages of these trade (Chinese ceramics) wares. Since there is insufficient land archaeology data available shipwreck data have been used to obtain probable dates of the Chinese ceramics. The invention of porcelain is China's outstanding achievement. Under the Tang period (618-906 AD) white porcelain was developed in China, but the blue and white porcelain came to limelight in the 14th century. The designs on the Chinese ceramics can be divided into six parts: animals and insects; flowers and fruits; figures and landscapes; and in *Kraak* style willow pattern and ending with formal designs and Chinese characters. The Chinese porcelain is made of a mixture of kaolin (China clay) and decomposed feldspar (*petuntse*) which is fired at 1300^0 C. Chinese porcelain have been reported from numerous shipwrecks. It appears that Europeans were unaware of Chinese ceramics until the Portuguese began to sail to the Far East in the early 16th century (Sassoon, 1991). The appearance and usefulness of Chinese ceramics were admired in the whole of Europe. Recognising the considerable profit made by the Portuguese and Spain in ceramic trade the British and Dutch also involved in this venture. Chinese porcelain had begun to arrive in England at least two decades before the establishment of the East India Company.

The bases of glass bottle indicate that olive oil, spirit and wine are imported from Portugal to Goa. Sometimes bottles contain alcoholic, non- alcoholic drinks, chemicals, medicines and many other materials (Orser, 2002). Even, the square glass bottle bases resemble the glass bases found in *Santo António de Tanna* 1697. Pontil marks can be seen on the bases of the square glass bottle which are unfinished whereas no marks are present on round bases. As may be expected, bottles were produced in many shapes, sizes and colours. The hippopotamus teeth and elephant tusks are imported from Mozambique and other African countries to Goa, then sent to Gujarat for carving sculptures and making handles, chess pieces, dices, etc. finally the finished artefacts were sold in Portugal and her colonies. The marks on the ivory could be the dealer's mark. There was much demand for ivory for image carvings for the churches and domestic use. Hippopotamus teeth were apparently in demand in the Far East. The preference for African ivory seems to be because of malleable nature, unlike the Indian ivory. Like Sunchi Reef many other shipwrecks have brought in hippopotamus teeth and elephant tusks. For instance, hippopotamus teeth have also been reported from the Uluburun shipwreck off Turkey coast datable to the 14th century BC (Bass, 1996). Recently, the exploration of a Portuguese shipwreck of 1530 at Oranjemund on the southern coast of Namibia brought to light large number of unworked elephant tusks. Similarly, the *Fredensborg* shipwreck of Danish-Norwegian origin has yielded 927 kg of elephant tusks and hippopotamus teeth of African origin (Chirikure et al., 2010;

and Delgado, 1997). The local divers and fishermen they say that had collected huge number of hippopotamus teeth and elephant tusks as well as Chinese ceramics from the site.

The size and shape of the iron anchor resemble the anchors kept in the Archaeological Museum at Old Goa and others found in Aguada Bay (Sila Tripati et al., 2005a; and Sila Tripati et al., 2009). Generally these iron anchors had the wooden stocks, however the anchor stocks of Sunchi Reef and Old Goa are missing. These anchors belong to the 'Admiralty Pattern Long Shanked' type and datable to the 16th-18th centuries (Sila Tripati et al., 2003a). Granite blocks are lying randomly on the seabed. It appears that these dressed granite blocks initially fell on the laterite boulders then slipped into the crevices of laterite. These blocks could be the main cargo of the ship and also served as ballast. The investigation has yielded a limited quantity of glass bottles, elephant tusks and hippopotamus teeth, pottery, Chinese ceramics; hence the mystery remains as to what could be the main cargo of the ship? It appears that the granite blocks could be the main cargo and the other findings are the part of secondary cargo. Because studies show that granite blocks were not only transported on ships as ballast but also as building materials. These granite blocks might have been brought to Goa for various constructions of buildings, public monuments, sculptures, decorative objects and art pieces, flooring and construction of steps of various churches of Goa. On the basis of the weight of the granite blocks the carrying capacity of the ship could be around 600 tons. Instances show that since ancient days stone blocks were carried onboard as cargo. The explorations of *Carry-Le-Rouet* shipwreck near Marseille, France in 1984-85 brought to light cargo of stones datable to the 1st century BC which were acquired from a nearby quarry. It is believed that the cargo was probably intended for building city walls. Since Greek quarryman marks were found on excavated stones which are similar to those used on the Hellenistic wall of Marseilles (Guerout, 2002). Similarly, the *Porto Novo* shipwreck which was excavated in 1992 off the Island of Corsica carried marble blocks ranging from 2 to 5 m long and 90 cm high and four columns of 150 cm diameter and 2 to 5 m long, for a total weight of about 138 tons. A number shipwrecks have been located off western Mediterranean Sea carrying marbles datable to the 3rd century AD onwards (Guerout, 2002).

Very few stone and clay bricks have been recovered from Sunchi Reef. Some of them are rolled and broken; these were used for very specific purposes. Correspondingly, yellow bricks measuring 157-180, 78-95 and 35-45 mm have been recovered from the Dutch shipwrecks of the 16th-18th centuries. Every year the VOC used to send thousands of bricks to Asia which served as ballast for the ships. These bricks were made in kilns along the River Hollandse IJssel near the city Gouda. In The Netherlands these were called *IJsselstenen* or *Goudse stenen.* Similarly, yellow bricks have been reported from the *Avondster*

(1659) shipwreck off Galle harbour, Sri Lanka, which was used on flooring of the ship. The finding of yellow bricks at the wreck site often identifies the Dutch ship. These bricks were carried as ballast to Asia and used there for construction (Parthesius et al., 2003). Significant numbers of bricks have been reported from *Batavia* (1628) *Vergulde Draeck* (1656), *Slotter Hooge* (1724), *Adelaar* (1728) (Martin, 2005) and *Zeewijck* (1727) (Delgado, 1997; and Ingelman-Sundberg, 1976; 1977). Probably, further research on this wreck would be able to give some more information on the issues which are unknown now. However, on the basis of the radiocarbon date of ivory, thermoluminescence date (TL date) of storage jar (360± 40 years old) and date of the iron anchor indicates that the wreck could be dated to the early 17th century AD, wooden hulled ship and belongs to the Portuguese period. Obviously no timber has been found in the wreck site which might have been washed out subsequently after wreckage. The length, width, masts, decks and other details of the ship could not be gauged now. Moreover, evidence shows that Portuguese had wooden sailing crafts till the 19th century AD. The two other 17th century AD Portuguese shipwrecks such as *Nossa Senhora dos Martires*, off Lisbon in 1606 and *Santo António de Tanna* off Mombasa in 1697 were wooden hulled ships. Findings of these two Portuguese shipwrecks resemble Sunchi Reef. Furthermore, the shipwreck findings are the evidence of maritime trade between India and Portugal.

Dutch and Portuguese Naval warfare

Adil Shah of Bijapur had strong resentment against the Portuguese. In order to drive them away from Goa Adil Shah entered into an alliance with the Dutch in 1637. According to their agreement the military force of Adil Shah would attack Goa through land area while the Dutch would blockade the Goa bar. If they were successful then Goa and its neighbouring land would go to Adil Shah and the Dutch were to get a place and many privileges in the kingdom of Bijapur. The Dutch blockaded Goa bar which was in fact effective between 1636 and 1641. In 1639, the Dutch burnt three Portuguese warships at Marmagao harbour. Those were the best warships the Portuguese had and indeed their losses were heavy. It is estimated that between 1629 and 1635 Portuguese lost 1499 men and 155 ships in their confrontation with the Dutch in the East. The question arises why Sunchi Reef shipwreck could not be the Dutch shipwreck. The probable answer could be 'no' because during the 17th century several naval battles have been fought between the Portuguese and the Dutch either at the outer side of Aguada Bay or at the entrance of the River Mandovi, but not at Sunchi Reef. Even several times the Dutch blocked the Goa bar. Both have lost their ships in the naval battles. For instance the Dutch burnt three Portuguese warships at Marmagao harbour in 1639 and it was a big loss to Portuguese. Generally battles were fought at such locations where the attacker can flee after damaging the enemy ships, for this purpose Sunchi Reef is not at all an ideal place. Secondly, if it was a naval ship, it would have carried more quantity of armament rather than four guns, one iron shot and a handgun and not such a number of granite blocks. Further, if the ship wrecked in naval battle it would have sunk either off Aguada Bay or in Mandovi River not at Sunchi Reef. Ships cannot come from either of these sites to Sunchi Reef because the adjoining area is covered with outcrop of laterite boulders. During recent exploration, a highly eroded iron anchor, an iron rudder, timber extremely affected by teredo, some wine bottles including iron pieces are noticed in the adjoining area belonging to a shipwreck of later period. Local fishermen being enquired regarding these finds it was suggested that some years back a fishing trawler capsized in a stormy cyclone. These remains belonged to the fishing trawler which certainly affects dating the original shipwreck findings.

Conclusion

Due to non-availability of archival information, it is not possible to state whether the ship was built in India or not, all the crew onboard died or survived, as also its name, name of the captain etc. As no timber remains of the ship have so far been found, its construction pattern and other details remain obscure. The exact causes of the loss of the ship remain unanswered but the results of the survey leave a number of possibilities open, which can be taken into consideration. From the findings it indicates that the ship was a cargo vessel but not a naval ship and it could be one of the Portuguese ships wrecked in Goa waters. Sunchi Reef shipwreck is the oldest wreck so far surveyed in Indian waters and the first Portuguese wreck until now investigated. The wrecked ship of Sunchi Reef was wooden hulled cargo sailing ship. As the ship was wrecked in the rocky area survival of perishable and light material is ruled out. The waves and currents might have also washed out the timber and other parts of the ship and some antiquities. Except the granite blocks, hippopotamus teeth and elephant tusks, no other cargo has so far been found. The limited numbers of findings suggest that soon after the wreck some of the cargo and other valuables might have been salvaged and such references are recorded in the marine records. The Sunchi Reef shipwreck findings are indicative of the trade and commerce between Goa and Portuguese empire as mentioned in the literature. Shipwrecks prior to the Portuguese period have not been located in Goa waters; hence the present findings are breaking new ground in the field of nautical archaeology of India in general and Goa in particular. Further, the Sunchi Reef explorations have opened a wide scope to undertake explorations of unknown wrecks because Goa has a rich potential for shipwreck archaeology in India. If further investigations could shed new light on the origin and date of wreck these would be an additional step.

4

Exploration of Basel Mission Company Shipwreck off St George's Reef

Investigation of a shipwreck is an important component of maritime archaeology; and starts with research into the history of the ship. Shipwreck archaeology is a field of historical archaeology as a whole and contains evidences to maritime heritage. Shipwreck exploration provides information on maritime history and contacts, age of the wreck, cargo, destination, flag of the ship, etc. A shipwreck is a storehouse of knowledge of the history of shipping and navigation. Shipwreck study provides us to understand the technical achievements and interactions of people in the past. The underwater excavation of ships has revealed the design and construction details of watercraft that have been wrecked either in the oceans, seas and lakes. Subsequent to the wreck event parts of the ship, organic remains and cargo are washed away, while metal and other heavy substances are preserved on the seabed. Shallow water wrecks are found severely damaged and scattered over a large area after striking against rocks or reefs, highly effected by environment, encrustations and marine growth, ransacked by non-maritime archaeologists and decayed over the period of time. Moreover deep water shipwrecks are found intact (less damaged comparatively than shallow water wrecks), less disturbed, not ransacked, less covered by marine growth and encrustations and with artefacts better preserved. In order to understand the underwater heritage, cultural contacts and preservation of their remains many developed and developing countries have made considerable progress in maritime archaeological studies by locating and excavating shipwrecks. Regular and organised explorations of shipwrecks were carried out in Goa waters from 1997 onwards. In addition to exploration of shipwreck off Sunchi Reef, Grande

Island, Amee Shoals and waters off Baga, a shipwreck has been discovered and explored off St George's Reef.

Topography of St George's Reef

St George's Reef is located on the eastern side of Grande [Grandi] Island and to the south of Marmagao port (Fig. 31). St George's Reef, Grande and Pikene islands off Goa appear to be detached portions of coastal headlands, now isolated from the receding coast as a result of erosion or rise in sea level over the centuries. Like other islands of Goa even the present Marmagao headlands were once an island. They are now connected to the mainland by a peninsula on which Vasco-da-Gama town now stands (Wagle, 1982). The tidal and longshore currents accompanied by strong winds (which becomes more vigorous during Southwest monsoon) blowing from the Arabian Sea might have played a predominant role in shaping the entire coastline of Goa. Evidences of sea level changes in the form of occurrence of marine platforms and beach rocks in the coastal belt indicate that at one time the sea encroached on the plains of Goa. The entire surface of the St George's Reef can be seen clearly during the lowest neap tide. But the reef fully submerges during the highest spring tide and it has a long stretch of underwater extension on its southern side. The seabed topography around the reef comprises coarse sand, fine silt, varieties of shell and seaweed, soft corals, gorgonians and other marine life. The reef is a navigational hazard for the mariners. The seabed of the St George's Reef region is loose, hence artefacts have been observed protruding from it. St George's Reef slopes towards the south and south-east side. On account of the gentle slope on the south and south-east side of the reef the water depth varies from 13 to 15 m at the wreck site. Sometimes during neap tide currents followed by heavy winds churn the suspended sediment in the reef and reduce visibility which usually takes a week to settle down and thereby improves the visibility after which the underwater explorations can be resumed.

Fig. 31: General view of St George's Reef and Grande Island

Exploration of Shipwreck

Marine and other related shipping records that are housed in almost all archives in India form the primary sources of information on shipwrecks such as the location, cargo, destination and other related details. The Naval Hydrographic Survey and British Admiralty charts also provide evidence and information on the location of shipwrecks. Based on the available data, a search for wrecks in the most likely area was initiated. The help of local divers was crucial in locating the shipwreck of St George's Reef. Sometimes information of fishermen is also highly critical for locating shipwrecks. Circular and swim line search patterns were followed for undertaking exploration of shipwreck off St George's Reef. Each diving team covered a circle of about 5 to 10 m radius in a circular search pattern and each dive sometimes lasted for more than 30 minutes duration. The water depths at the dive site varied from 13 to 15 m. During the course of exploration, attention was paid to the occurrence of natural features such as outcrops of rock, seabed topography, water depth, current direction and visibility at the site. The entire area was covered comprehensively in order to understand the distribution, state of artefacts and extent of the wreck site. In the course of explorations, the smaller and important artefacts suitable for retrieval were collected for study and analysis.

As the artefacts of the shipwreck are either buried and protruding in the silt it was decided to undertake probing in the area to gauge the extent of the wreck site, state of the artefacts and depth of burial in the sediment. In order to carry out the probing one meter long 'T'- shaped iron rods were used to probe the site from which the nature of the seabed around reef is inferred. During probing, utmost care was taken to prevent possible damage to buried archaeological objects. To cover the whole site and for better exploration 'T' pegs were fixed from a base point on the east, south and south-western flanks and the whole area was connected with polypropylene rope and positions of 'T' pegs were carefully checked and charted through a handheld GPS. Surveys using Pulse 12x boat-towed and Pulse 8x hand-held metal detectors were carried out in all directions within the wreck site to record metallic anomalies. This is because, after the wreck, all the part of the ship and cargo are expected to be scattered over a vast region, subsequently, some cargo would be buried under sediment and some washed out. Underwater excavations are carried out either by airlift or water dredge, which are used to remove sediment deposited on the artefacts by suction. In some places hand fanning was followed up to expose the buried antiquities in the sediments and airlifting was carried out up to 1 m seabed depth to ascertain the stratigraphy of the seabed and to find out how deeply artefacts were embedded. During airlift variety of shells, barnacles, urchins, dead mussels and compact sticky clay and more number of broken roofing tiles and flooring tiles were observed. After ascertaining the distribution of artefacts positions were taken using GPS (Magellan model NAV 5000 DLX), followed by systematic underwater exploration, documentation, recording and excavation including the environment of the site.

Shipwreck Finds

Visual exploration revealed that the remains of the shipwreck lie scattered over a larger area with the highest number on the south and south-east side of the reef, because of the sloppy seabed. A large number of terracotta artefacts such as furnace (chimney) bricks (Fig. 32a), floor tiles (Fig. 32b), roof and ridge tiles (Fig. 32c), column capital and hollow column drum (Fig. 32d) and drainage pipes (Fig. 32e) as well as the timber of the ship found in 15 m of water (Sila Tripati et al., 2002). Due to the cleanness of this region's water, biological growth such as coral, orange-coloured sponge gorgonians and other marine lives are notable. Though both Pulse 12x boat-towed and Pulse 8x hand-held metal detectors were deployed to ascertain the presence of ferrous and other metallic objects and artefacts either buried or lie on the seabed but the survey did not reveal any metal anomalies, thereby indicating that the ship was probably a country craft built of wood. Tiles and other artefacts had turned black due to long burial in the seabed.

Furnace Bricks

Bricks are made of white clay and are used for the chimneys in surviving houses and furnaces of this period. Probably, these bricks are used for industrial purposes. On each brick the words '*Basel Mission Company 1865*' are stamped only on one side where as the other side is plain and the stamped side of the bricks has a frog. Bricks measure 230 X 115 X 60 mm, each brick weighs 2300 gm. Microscopic examination of a brick section was carried out. A thin section showed a variety of grains among which quartz is predominant. Most of the grains are angular to sub-angular in shape. The size of grains varies from fine to coarse; fine grain comprises less than 20 %, whereas coarse grains comprise 70 to 80 % of the total. Other brown and opaque minerals observed in very negligible quantity [1 to 2 %] are cementing material. Some portion of the section shows red coloration. Cementing material is also siliceous. Megascopic study showed that the brick consists of coarse quartz grains (angular to sub angular in shape). The grains show a variety of colours which are predominantly white and a shade of pink. They are bound together by cementing material, which is white in colour. On the basis of these analyses it can be inferred that the bricks were made of local sands derived from acid plutonic rocks such as granite, granite gneiss and quartz veins. The material has not been subjected to lengthy transportation, as is

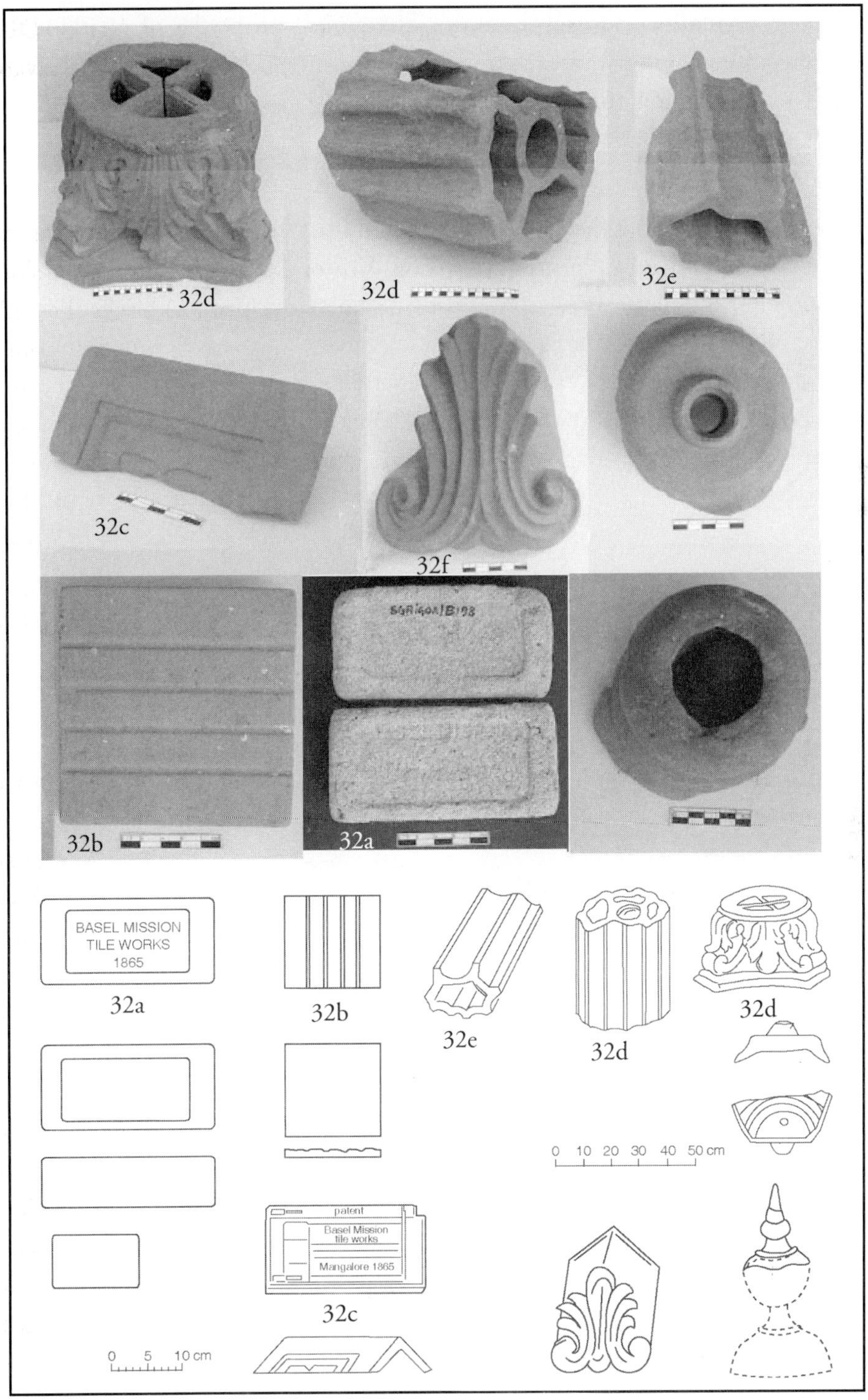

Fig. 32: (a-e) Types of tiles and other artefacts of the Basel Mission Company

evident from their angularity and coarse nature. The optical properties of quartz grains point to their derivation from quartzite rocks. The nature of the cementing material can be confirmed by X-ray diffraction analysis.

Floor Tiles

The floor tiles are square in shape, neatly trimmed, uniform in thickness with four deep grooves at regular distances and are stamped '*Basel Mission Company 1865*' in circular design at one corner of the tiles. These grooves provide a better grip; hence this side was to be placed on the sub-floor. These are made of well levigated clay and well fired. Floor tiles are numerous than the bricks, roof tiles and other terracotta artefacts. Some of the tiles have turned black because of burial in the seabed for long duration. Each tile measures 220 × 220 × 20 mm.

Roof Tiles

The words '*Basel Mission Company 1865*' are stamped on the lower side of all the roof tiles. These tiles are lighter than other artefacts and hence breakage is more. Each tile measures 400 X 230 mm. Like floor tiles the roof tiles are made of well levigated clay and well burnt. In terms of breakage, more roof tiles are broken than floor tiles. These tiles are quite similar to other tiles in terms of look and size. None of the bricks and tiles has decomposed or disintegrated in spite of lengthy submergence in marine water, which clearly indicates the well levigated clay and firing.

Corinthian Column capital and Column drum

The capital is hollow inside and one end is closed. It is made on Corinthian design. Four supports were fixed inside it to provide additional strength and appear to intersect. Prolific marine growth was noted on the outer surface. The upper end of the capital, which would rest on a fluted column, is circular while the base on which the lintel was to be placed is square. At some places, a small chip has come out while lifting. On fitting the square portion remains on the upper side fixed with lintel whereas the circular portion is placed on column drum. The height, width and circumference of the column drum are 320, 300 and 930 mm respectively. It is fluted, hollow and partly broken. It was photographed *in situ* before retrieval for further study. These hollow drums are placed one on the above to make a column or pillar on which the column capital is placed. After placing one on the above these are joined with some binding material and finishing touch is given at last. Only one column capital and shaft have been recovered from the site.

Timber of Wrecked Ship

A single timber was found during exploration of the shipwreck site. Its maximum length is 2 m, width 180 mm and height 160 mm (Fig. 33) (Sila Tripati, 1999). It is 'U' form shaped and longitudinal rabbets suggest that it is part of the keel of the ship. The rabbets are 50 mm deep and the hog between is 100 mm wide. One end is well preserved and has the remains of a carefully worked scarf with in cuts from a chisel and nail holes. There are two nails on the lower side suggesting that a false keel or sheathing was fitted to it. Woodborers have destroyed the other end. Chisel marks are visible on the timber. This timber survived because it was buried in sediment within the close vicinity of the reef, only little portion was visible on the seabed and was not affected by current action. From the timber it is possible to infer that the ship was not large and had a wooden hull with 50 mm (2 inch) planking. It is assumed that the shipwreck broke up and all the other wood and organic material might have been washed away, decomposed or destroyed by teredo. A small piece was collected from the timber for radiocarbon dating and anatomical analysis (Sila Tripati et al., 2005a). The timber is highly deteriorated and affected by woodborers; hence growth rate cannot be identified. The result of radiocarbon dating of the timber is 114.3 ± 1.5% years old (measured C^{14} Age $T^{1/2}$ = 5570 ± 30 years) and laboratory sample No. is: BS-1582. The timber is incomplete, which prevents to know the overall size determination. The anatomical analysis shows that the timber belongs to the *Lagerstroemia Lancealata* species, for which the trade name is benteak (for detailed analysis see pages: 163-171).

Discussion

Several ships are known to have sunk in Goa waters but not all of these are marked on Hydrographic charts or mentioned in marine records kept in Goa State and other Archives in India and abroad. Moreover, the shipwrecks which are marked on Hydrographic charts do not indicate their nationality and dates are not mentioned and it is also not possible. Even, the St George's Reef shipwreck is also not mentioned in any records (Sila Tripati et al., 2003b). The discovery of the wreck has elucidated the history of the *Basel Mission Company* and its trading activities in the past two centuries. The shipwreck could have been caused when the ship struck the shallow submerged reef, as a result of human error in navigation or inclement weather. One of the main reasons for the shipwreck in shallow water was inadequate knowledge about sea bottom topography. The concentration of artefacts on the south and south-eastern side of the reef suggests that the ill-fated vessel was coming from the southern side of Goa (Mangalore or Calicut) when it struck the reef and foundered. Exploration yielded only one column capital, drum, ridge tile and basin whereas

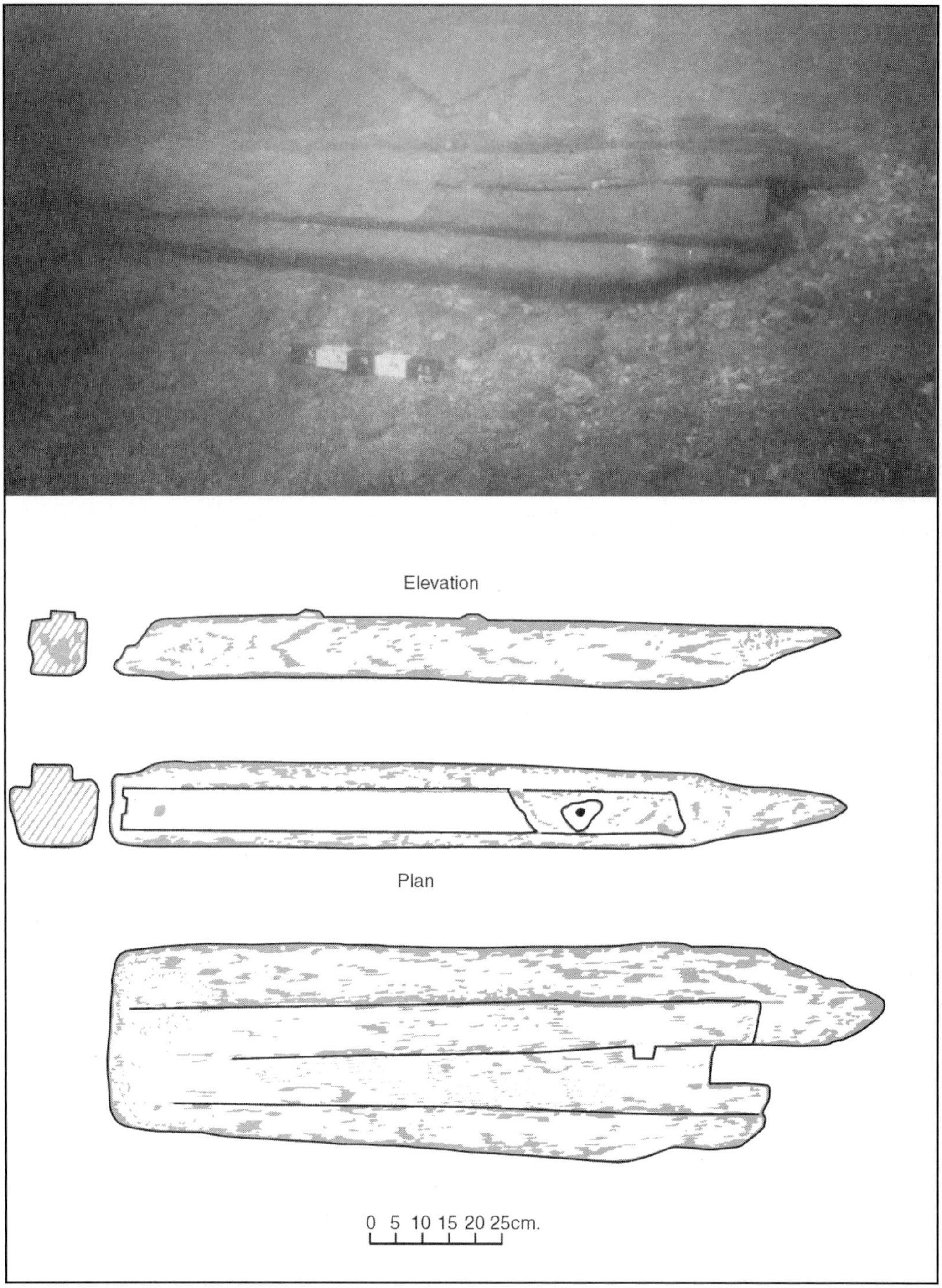

Fig. 33: Remains of timber from the wrecked vessel on St George's Reef (Scale in 5 cm units)

bricks and roof tiles were more in number, because tiles and bricks are used in larger quantity in house construction. These artefacts are buried in the sediment. The quantity of artefacts lying as well as buried in the seabed might be the full consignment for a country craft. However, the exploration is confined to a limited area hence a capital and a column drum were found, but more number of these artefacts could be expected. There could be other possibilities that some of the column capital, drum or shaft and ridge tile might have been lifted by the local divers as they are unbroken and could be reused after removal of marine growth. The other possibilities could be that soon after the wreck other terracotta artefacts might have gone into the seabed because of the gravitational force or movement of sediment, seasonal monsoon and cyclones might have buried them in the seabed. In view of this extensive airlifting was carried out. In the course of airlifting, bigger pieces of tiles were noticed in the upper levels whereas smaller pieces were found in the lower level of the seabed. The finding of Corinthian column capital and column drum suggests that the craftsman of the *Basel Mission company* were aware of the Hellenistic and Roman styles. It appears that the ship was loaded with only terracotta artefacts intended for house construction.

The column drum and capital are hollow inside because it helps in uniform firing and not to be cracked under shrinking stresses in firing, it is also easy for handling and light to transport, less clay required as it is hollow. Further, during construction wooden posts or iron rods were provided inside the hollow of the drums to provide additional strength to the columns and to prevent from moving the drums from its original position. The roof and ridge tiles are light and elegant in appearance. In addition they are cheaper, durable, more easily laid than conventional tiles due to their double interlocking design. When properly fitted they make a long lasting perfectly wind resistant and watertight roof. To cover an area of 30 to 35 sq m 130 to140 tiles are required depending upon the spacing. Each ridge tile covers two flat tiles on each side of a pitched roof. The tiles are turned out by machinery to achieve a high level of strength, finish, overlap and catch groove. Floor tiles are plain and destined for houses, offices, warehouses and factories. The other household decorative terracotta artefacts are comparatively less than bricks and tiles. The flooring square tiles are more in number and broken into two pieces whereas roof tiles are less and broken into a number of pieces. The majority of bricks are intact. A large number of bricks, floor and roof tiles are noticed near the timber. Comparatively, the bricks are less in number than the tiles. These tiles were in great demand as inexpensive hard-wearing material for flooring. All the artefacts are made from well-levigated clay and well fired. None of these artefacts have soot marks which show the uniformity in firing and constant maintenance of firing temperature. They do not show any spalling, cracking or other defects despite lengthy burial in a marine environment, except marine growth, this shows the quality of the artefacts. Soon after removal from the artefacts from the seabed those were washed in running tap for several times and kept in fresh water for some days. Subsequently the water was replaced and this practice continued for long time, then dried in shade without exposing to direct sunlight. After study and documentation those were stored in the storage room. However, remains of marine growth can be seen on them. For this reason, our measurements may not be precisely identical to the factory standards. Although the stamp *Basel Mission Company* was not found on all the artefacts, it is presumed that all were the products of this company but it is difficult to know why all artefacts are not stamped. Use of similar artefacts can still be seen in old Portuguese houses in Goa and coastal region of Karnataka and Maharashtra. The keel timber found on the site indicates that the ship did strike the reef and wrecked. The reef would be traversed by sail close to the coast and the wreck could have been caused at night and due to navigational error. The reef is dangerous for sailors who are not aware of the region. All the timber of the ship and other light materials i.e. cargo might have washed out because the reef region falls under high energy zone and subsequently might have disintegrated. Manual exploration and metal detector survey did not reveal any metallic anomalies such as a propeller, boiler, or any other metal artefacts so the vessel appears to a wooden hulled sailing ship (Sila Tripati et al., 2003b).

The Basel Mission Co. and its Impact on Society and Culture

Since the cargo belongs to the *Basel Mission Company*, it is important to trace its origin and other activities. German missionaries from Basel, border area between Germany and Switzerland established the company in 1834 in coastal region of Karnataka. Along with their missionary activities, they also ventured in industrial activities. The mission had 24 establishments in Karnataka, Kerala and Tamil Nadu. The *Basel Mission Co.* set up a printing press in 1841 and a weaving industry in 1844. Further, they established this first tile factory at Jeppo, Mangalore in 1865 and subsequently at Calicut, Kudroli, Malpe, Codacal, Palghat and Feroke (Fig. 34). Easy access of raw material and transport facilities made the company set up tile factories in these areas. This was their first industrial venture on the Malabar Coast of India (Raghaviah, 1990). Initially, bullock carts transported the tiles; subsequently boats were used for transportation (Hofmann, 1913), the timber which was found near the reef provides additional support to this statement. As manual exploration and metal detector survey did not reveal any metallic anomalies, it appears that the ship could be a local sailing craft. The manufacture of tiles was a centuries old occupation of potters in southern India. The Portuguese traveller, Duarte Barbosa who visited Malabar coast between 1510 and 1531 made the following observation on potters, 'Their business is to work at baked clay

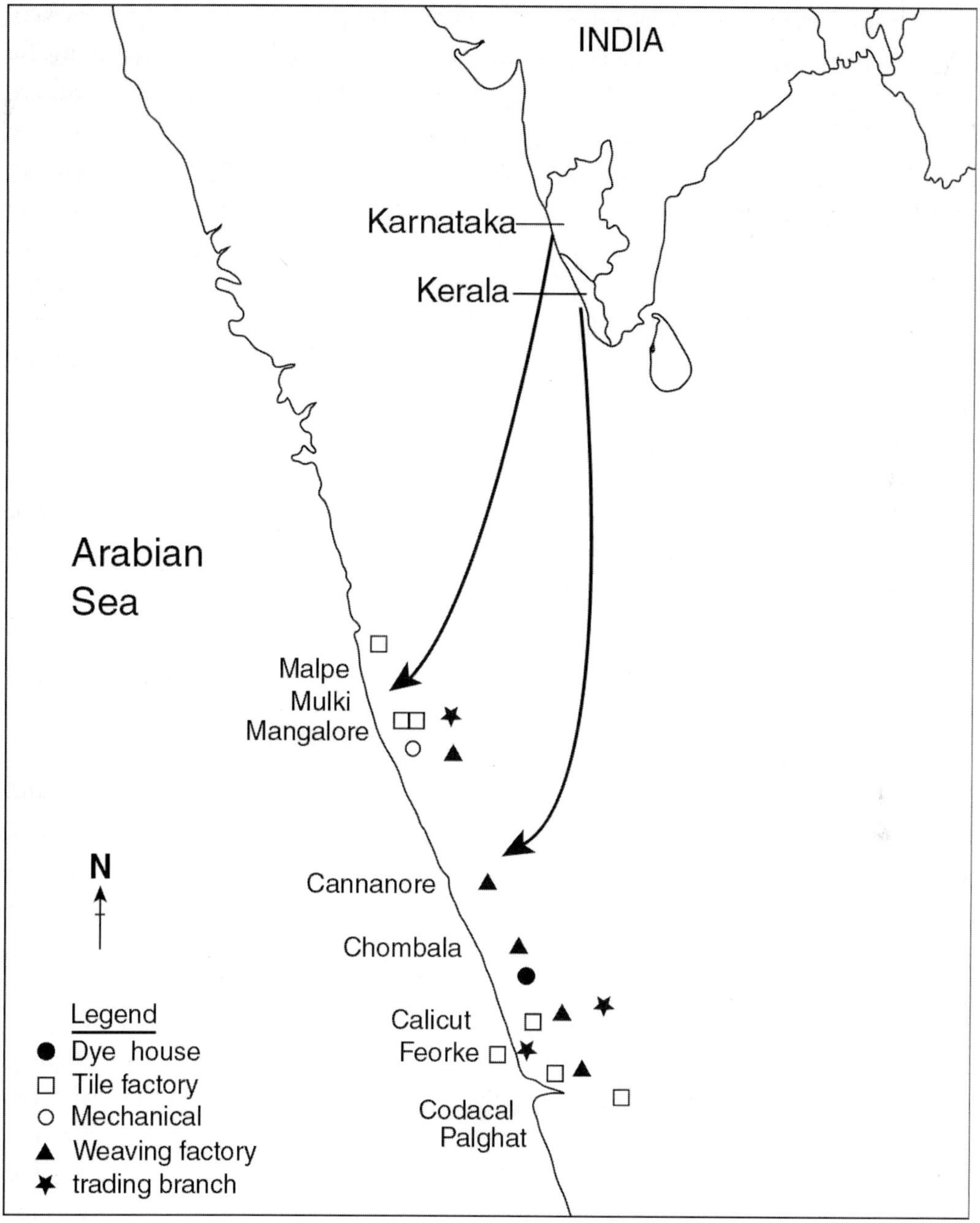

Fig. 34: Map showing the location of Basel Mission's industrial activities along the Karnataka and Kerala coasts of India

and tiles covering houses with which the temples and royal buildings are roofed and by law no other person may roof their houses except with palm branches' (Barbosa, 1970).

George Plebst studied the technique of tile making, including treatment of clay, glazing and construction of the kiln and firing process. The raw material for manufacture of tiles, namely clay or feldspar was abundantly available on the banks of the River Netravati in Mangalore, Karnataka. The company introduced further technical changes in the manufacture of tiles; these included salt glazes and gas-fired kilns in order to obtain uniform temperatures (Bliss, 1890). Subsequently, the company manufactured ridge tiles, both plain and ornamental, skylights and ventilators, ridge and hip terminals, finials of various kind, hanging wall tiles, ceiling tiles of many different designs, *hourdis* or ceiling slabs, hollow bricks, plain and ornamental floor tiles, chimney bricks, slat-glazed stoneware, earthenware, drainage pipes, terracotta vases, flower vases and architectural terracotta items. The *Basel Mission Company* introduced a fully mechanised system of manufacturing tiles, extending to mixing of clay, pressing to shape, cutting blocks, drying and firing, all of which was in earlier times done manually by the potters. Tiles were exported throughout India, British East Africa, Aden, Basra, Sumatra, Borneo and Australia (Hofmann, 1913). They were also exported to German colonies in Africa as is revealed by a statement by the agent of the Basel Mission (Raghaviah, 1990). From the entrepot Bombay (Mumbai) they were exported to other parts of the country. The company also operated in other parts of the world, namely Ghana, Borneo and China, but their industrial base remained strongest on the western coast of India because of low investment costs and ready availability of manpower.

The Church-oriented Pietistic Movement was established in 1780. In Basel a society named the German Christian Society was formed with the objective of restructuring the Christianity of the world. Basel was selected as the headquarters of the Society. Basel was located on one of the main trade routes connecting France, Belgium and Germany. The Basel Mission was founded in 1816 to train missionaries and it was supported by Basel-based members. The Basel mission started its work from 1834 onwards at Mangalore, west coast of India. The Mission stressed that every convert would be educated. Further, the converts were asked to provide labour for industrial activity. The Basel Mission established industrial enterprises in Malabar and south Canara for the converts. The economic activity of the Basel Mission is noticed in Malabar and South Canara region of Kerala and Karnataka. The Church records of Codacal Parish in south Malabar state that the converts were predominantly from lower castes. The mission was even succeeded in converting Muslims in small numbers. Apart from achieving social mobility the industrial labour had the added advantage of lessening social stratification among the converts. The general atmosphere of the Mission industries where converts from lower and higher castes worked side by side helped in the eradication of caste prejudices. Moreover, the Basel missionaries took initiative in arranging marriages between converts from different castes. The industrial activity of the Basel Mission can be divided into three phases. The early phase (1834 -1852 AD) begins with the arrival of missionaries in the Malabar Coast in 1834. Industrial activities, mainly local crafts started from 1846 onwards. In the middle phase the industrial activities increased. Handloom, weaving, printing press, and mechanical workshops were established in the Mangalore region. This period witnessed the trading activity of the Basel Mission. By 1870, the mission faced local competition due to establishment of new tile factories by others. Consequently, the company started large tile factories. In 1846, Industrial Commission was established by the mission with a view to centrally control the industrial activities. During the last phase (1882-1914 AD), the industrial commission was amalgamated with the missionaries who promoted higher capital investment and expansion of industrial activities on the Malabar Coast. However, in 1914 with the outbreak of the First World War, the industrial activity came to an end. In 1919 as fallout of the First World War the Basel Mission industries were taken over by the British Government. A new company, the Commonwealth Trust, was incorporated to look after the commercial ventures not only in India but also in coastal Africa. In 1977 the company passed into Indian hands, governed by an Indian Board of Directors and the Commonwealth Trust (India) Ltd. or Comtrust came into being. Over the years Comtrust witnessed two World Wars and both European and Indian rule. What remains intact is Comtrust, which has been a socially responsible company from its origins in the mid-19th century. Although the company retains no specimens of products made in the 19th century, there is ample historical evidence for Basel Mission and its manufacturers. The period 1865 witnessed a change in the manufacturing of tiles and other artefacts in the western India. This change had a strong impact on the traditional potters who suffered as a result. Today there are five Comtrust factories producing over a lakh of tiles a day and Comtrust tiles adorn some of the best-known landmarks in India and abroad. Comtrust has been following the same methods in selecting and sieving of clay as the *Basel Mission Company* had adopted and still today the converted (both men and women) are working there. Moreover, Comtrust has maintained the same quality and standard of production. Today, there is little demand for terracotta columns Comtrust is involved in producing only tiles, bricks, flower pots, vases, ceiling slabs and ventilators.

Conclusion

The impact made by Basel Mission industrial efforts in terms of transfer of techniques is undeniable. This resulted in the systematic application of new technical knowledge to indigenous industrial activities. The systematic offshore exploration of St George's Reef has brought to light an undocumented wreck of the company and suggests that it was transporting its products in country wooden ships. It seems that the ill-fated ship was coming somewhere from Karnataka or Kerala to Goa or Bombay (Mumbai). This is the first shipwreck of the company located during underwater excavations and dates to its period of expansion in the mid-19th century. Further investigations would provide more information on the Basel Mission and shipwreck. Similarly, underwater searches in Karnataka and Kerala waters may bring to limelight some more shipwrecks if any belonging to the Basel Mission Co.

5

Steam Engine Shipwreck Exploration off Amee Shoals

Over the period of time several changes have been observed in the shipping and shipbuilding technology. These changes have come into existence once ships carried more cargo, sailed to distant lands in order to meet the demands of mariners. More importantly, in the 19th century many advances were noticed in shipbuilding technology including steam power and iron hulls in place of wooden hulls and sails. The broad structural details of wooden ships are same as those of iron or steel ships, but the methods of construction differed because of the difference in the nature of the material. With the introduction of steam engine ships, the age old sailing ships ceased, and a new chapter was added to the shipping and shipbuilding technology. Steam engine ships could carry more cargo and move faster than the sailing ships. Information on wooden hulled and steam engine shipwrecks in Indian waters has been collected from archival sources but only a few shipwrecks have been explored. This is the first discovery of a steam engine shipwreck in Amee Shoals off Goa waters and its exploration becomes important in order to study the causes of the wreck, origin of the ship, the state of finds; the probable period of the wreck and reconstruction of the maritime trade contacts of Goa.

Topography of Amee Shoals

Amee shoals lies about 1.5 km south of Sunchi Reef and close to Marmagao bay. These shoals lie in North-South direction with a maximum water depth of 9 m and separates Marmagao bay from the Arabian Sea (Fig. 35). The height of the shoals varies from 4 to 6

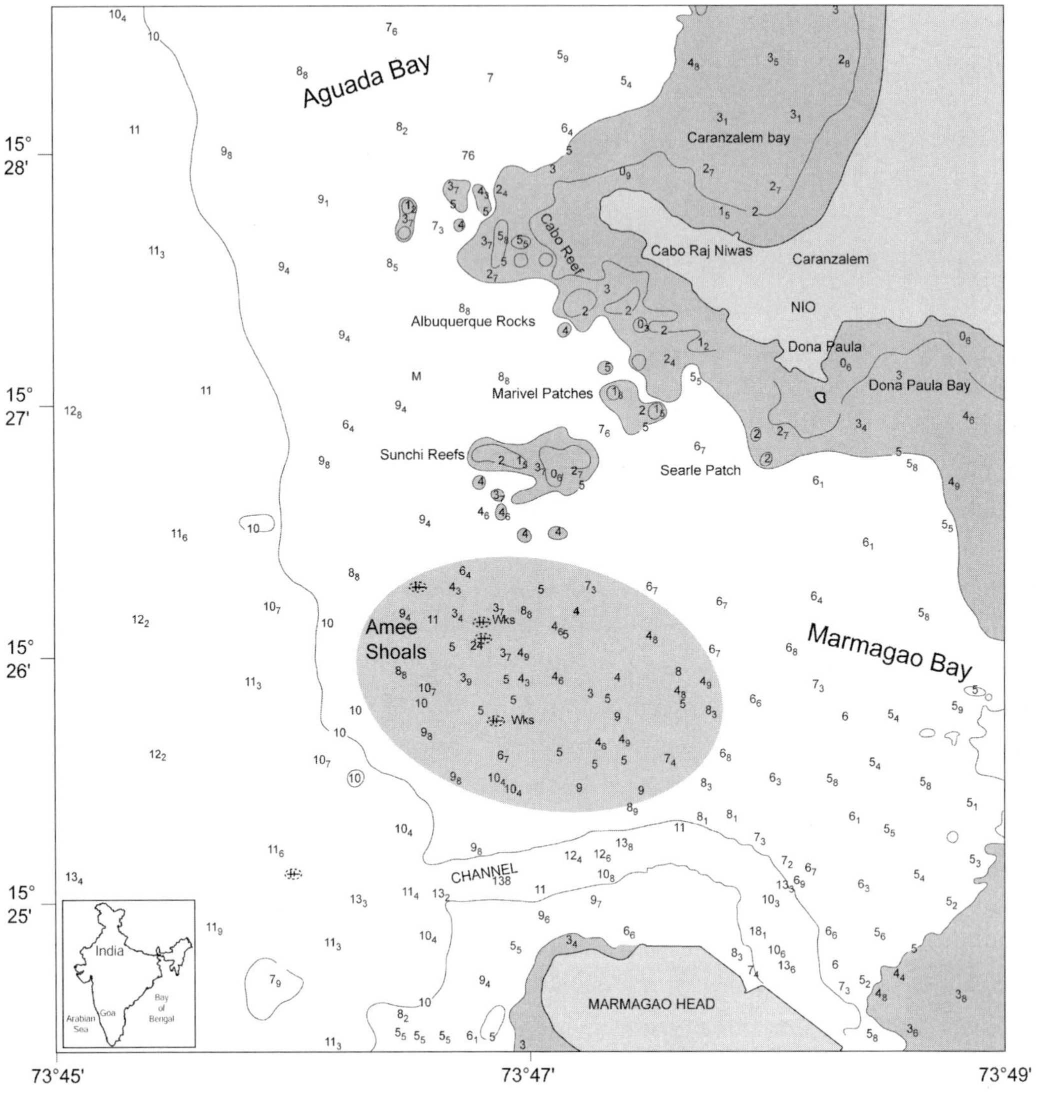

Fig. 35: Location of Amee Shoals shipwreck

m. The seabed of the region is full of coarse sand mixed with silt. Bathymetry of Amee shoals indicates that the area has a gradual slope towards the west. The shoals have a wider extension but are truncated at places with varied depths (Ambre, 1991). Growth of soft corals, small marine plants and gorgonian is noticed on the shoal and some places on the shipwreck. Crevices are also noticed on the laterite. The River Zuari discharges sediments into the sea near Amee shoals. Because of the suspended materials sometimes sunlight does not reach the seabed. The coarse material predominantly clay along with calcareous sand does not allow preservation of archaeological artefacts because this environment is highly susceptible to biofouling. The sea bottom in this area is covered with shelly sand, lithified sediment or a rocky bed thinly covered with sediment (Wagle et al., 1988). The current action at Amee shoals during the ebb tide is quite strong and causes difficulty for diver-archaeologists to undertake the exploration. The tidal current observed in Amee shoals is about 1. 05 to 1.08 knots during spring tide. The sea breaks heavily along this track and this region falls under high energy area. During the monsoon period the height of the waves in Goa goes up to 6 m while it is less than 1.5 m in fair season (SanilKumar et al., 2006).

Exploration of Shipwreck

Based on information available on Amee shoals from the Hydrographic Chart No. 2020 exploration of the shipwreck was carried out in 2008. Local divers were also consulted to get additional information on the site as well as shipwreck of Amee shoals. The position of the shipwrecks shown in the Hydrographic chart are marked in the GPS and navigated to the site with GPS. After reaching the site, with help of rubber boat (Gemini) on the basis of the GPS position a marker buoy dropped at the wreck site. After locating the shipwreck, underwater explorations were carried out at Amee shoals. The water depth varies from place to place in Amee shoals. In addition to underwater exploration, natural features such as outcrop rock, seabed topography, water depth, movement of sediments and current direction in the region were noted during exploration. Scuba diving is quite suitable for shipwreck exploration hence scuba diving was initiated and swim line and circular search patterns were followed. Each dive covered a large area of about 5 to 7 m radius. At places, hand fanning was carried out to expose the buried artefacts, if any, and parts of the ship in the sediment. After locating the remains of the shipwreck, all the finds were interconnected with rope to understand their distribution and positions. The boiler, boiler bricks, flying wheel and other objects were recorded in the Nikon RS underwater still camera to the possible extent and the co-ordinates of the explored area and shipwreck findings were taken by Magellan model NAV 5000 DLX GPS. During underwater survey information on outcrop rock features, seabed topography, water depth and current direction in the region were also obtained.

Shipwreck Finds

In the course of exploration boilers, boiler bricks (furnace bricks), flanges, broken copper pipes and tubes were noticed which are scattered over a wide area. Three boilers, flywheel, Hawser and Hatch door and water tank of the wrecked ship were found. The other parts of the ship namely cabin; deck and hull were either ransacked or corroded. In order to study and reconstruct the history of the ill-fated wreck a few boiler bricks, flanges and copper pipes were collected. The local divers have ransacked all the brass parts of the ill-fated ship (Sila Tripati et al., 2010a).

Boilers

Three boilers were noticed during exploration which lies close to each other in north-south direction. The rear side of the boilers faces the southern side. They are made of wrought iron with single riveted lap joint and their size, shape and diameter are same with a height approximately 4 m. As brass and copper has been used extensively in different parts of the boilers the local divers have despoiled the boilers in recent past. The main steam stop valves of all the three boilers have been ransacked by the local divers due to their highly valuable non-ferrous fittings (Fig. 36). Each boiler has 3 furnaces and a row of heat exchanger tubes and stay tubes are fitted. The fire tubes are at the end of the boiler, while the combustion chamber is on the other end. Manholes and safety valves of the boilers have remained intact. The steam out pipe is present on the top of each boiler. Only one furnace of a boiler is intact with its blanking flange, while the other eight furnaces of three boilers have been destroyed, probably bricks and other parts have also been robbed. Various types of bricks with different sizes are placed

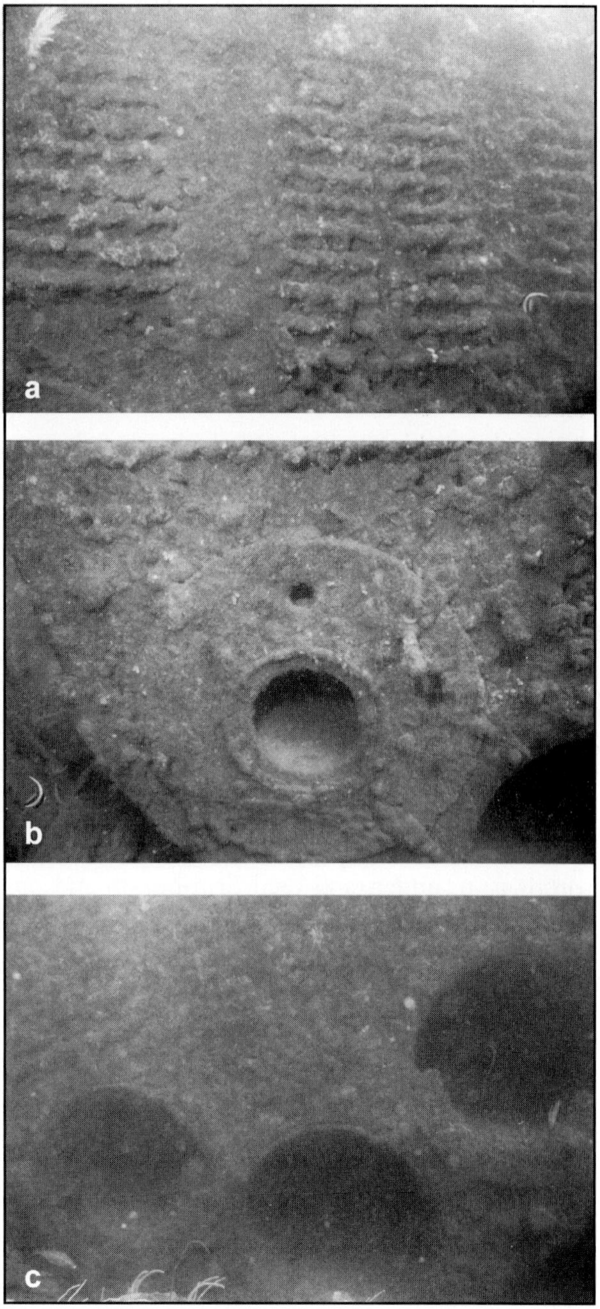

Fig. 36: Boilers of the steam engine shipwreck off the Amee Shoals, Goa

inside the furnaces in several layers in different directions. The first row of bricks is placed horizontally, then bricks of tapered shape are placed as the second row, above this another row of bricks are laid horizontally. The diameter of the furnace hole is made smaller, by providing a number of rows of bricks and then a metal flange is fixed on the front side of the furnace. Bricks are used to hold heat and protect iron from fire. Above these furnaces, thin coiled fire tubes are fitted. Stop steam valves are present on the boilers. On the upper side of these boilers there are two rows of projected objects which appear like knobs; on the first row there are four and five are on lower row. There is a circular opening at the middle. At the lower side a handle has been provided probably to release the steam?

Boiler bricks (Furnace bricks)

Boiler bricks were made of refractory material; generally mixed of silica and aluminium oxide with traces of iron, calcium and other oxides (Osbourne, 1941). Boiler bricks of Amee Shoals shipwreck are made of different shapes and thickness which could fit properly in the front, back and other sides of the boilers. These bricks are made in such a way that they develop their strength only after firing to a high temperature. Several boiler bricks are found scattered underwater around the boilers. Some of the bricks appear to be much heavier, longer and thicker (Fig. 37). These bricks are placed on the furnace front. The purpose of putting up bricks inside the furnace is to resist high heat and temperature as well to provide insulation. Unless bricks are provided furnaces might develop spalling and cracking because of heat. Bricks are positioned in horizontal and vertical manner inside furnace then the tapered shaped bricks are placed above them. To reduce the diameter of the furnace more layers of bricks are arranged. The joints between bricks are sealed by applying binding material. The binding material is presumably

Fig. 37: Boiler bricks collected from shipwreck site on Amee Shoals (Scale in 5 cm and 1 cm divisions)

made of equal quantities of fireclay and finely crushed firebrick mixed with water to form a mortar. Binding material is not visible underwater because a layer of silt is deposited on these bricks. After the completion of lying of bricks inside the furnace the flange is provided on its front side. In between the bricks and flange, a layer of asbestos millboard sheet is provided and spring washers are fitted under the nuts that screw on to the brick bolts. Some of the bricks which have been collected from the site have stamp marks. On the upper side of the bricks 'FURNA....' some words are illegible and in second line 'B 84' has been stamped. There are no markings on the other side of the bricks. Some bricks have been retrieved which are joined with binding material whereas others have depressions probably for joining with other bricks. The measurements show that they vary in size. Numerous bricks are still inside the boilers and some are trapped in between boilers and hull of the ship. Marine growth and deposit of silt were noticed on the boiler bricks (Sila Tripati, 2008). The 'FURNA.....' stamp on the bricks is indicative of the type of materials used for furnace linings. The details of 'FURNA.....' brick manufacturing company is awaited, more light can be shed on the shipwreck and boilers once the information is obtained.

Flanges

Flanges are very common on steamships. Three types of flanges belonging to different parts of the steam system were collected from the site. One among them is a circular flange having eight holes (Fig. 38a). The circular flange is provided on the front side of the furnace of the boiler. The diameter of the flange is 52 cm and distance between the holes is 18 to 20 cm. Stamps or any other marks are not visible on these flanges. The fitting of flange indicate that these are fitted for oil burning but not coal. During survey, similar types of some more flanges were noticed which are trapped between the boilers and hull frames. Marine growth is present on these flanges. In another flange six holes were noticed and its lower portion is sealed. Two stamp marks are visible on it, one reads

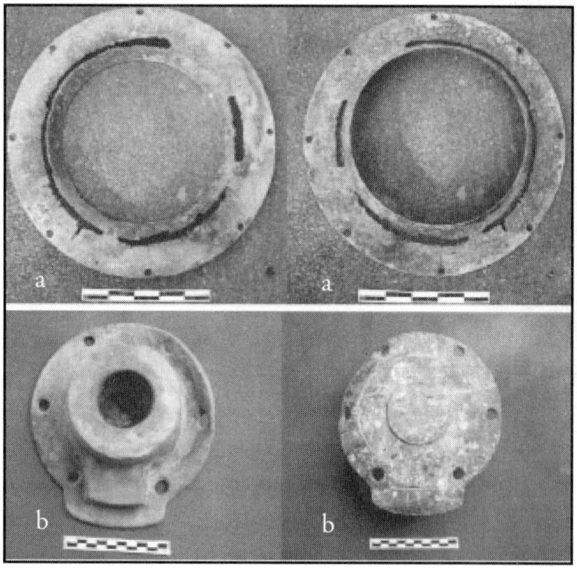

Fig. 38: Flanges recovered from the wreck site on the Amee Shoals

'SOUNDING' and the other reads 'No. 3 TANK' (Fig. 38b). The Tank No 3 sounding is for dipping a stick or metal rod with graduations in it for checking the fluid levels in the tank which is often called 'sounding'. This flange could be from an oil tank. Before the introduction of gauges these types of flanges were common and are still used today as a double check on gauges. The other flange is oval shaped having two holes at both sides and the centre portion is elevated in which a tube is connected but no markings were found on it.

Water Tank

The standard large metal water tank measuring 4 x 1.65 m is noticed lying close to the boilers. This water tank is not pressurized, but probably used for recirculation of water from the condenser. The tank has been reinforced with steel frame then plates are provided from all four sides (Fig. 39). This type of water tanks were a common feature in steam

Fig. 39: Water tank of the wrecked ship lying near the boilers on the Amee Shoals

engine ships. Two pipes are fitted horizontally inside the water tank. The tank is highly corroded and marine growth has taken place on it. Fresh water was fed into the boilers from the water tank where it is heated and converted to steam. The steam was fed into the engine directly to the high pressure cylinder. The steam was exhausted from the HP cylinder into the MP cylinder and in turn to the LP cylinder (cylinder diameters increasing from HP to LP). Once the exhaust valve opened on the LP cylinder, the steam was sucked into the condensers due to the vacuum effect of the condensing steam. The steam condenses to fall as water to the bottom of the condenser, where it could be fed back into the boilers.

Hawser and Hatch Door

The bow section of the ship has been located and this appears to be the Hawser (Fig. 40), for anchoring the ship. It has a large lip edge. Obviously, there should be two of these

Fig. 40: Hawser for the anchor of the wrecked ship (Scale in 5 cm divisions)

hawsers for the port and starboard bow sides. However, we have located only one Hawser. The other one is yet to be located. Similarly, the hatch door of the ill-fated ship has been found in the adjoining area of the boilers. The iron hatch door (Fig. 41) is the part of the steam drive system, which is used for an inspection and heavily built to withstand heat. Two iron rods, each 85 cm in length have been joined at the inner side of the lid to provide reinforcement (Sila Tripati, 2008).

Fig. 41: Hatch door of the wrecked ship noticed from Amee Shoals shipwreck

Flywheel

The flywheel of the ship is lying on the northern side of the boilers. Some parts of the flywheel were found lying on the seabed close to each other. One of them is buried in the sediment and overgrown with marine organism. The maximum length and width of the flywheel is 180 and 60 cm. The flywheel of a steam engine ship runs with a constant angular speed or rotational velocity. When the steam is shut off, the friction of the bearings and the air stops the flywheel to rest.

Discussion

The study of steam engine shipwrecks is important in terms of their development and technological aspects. The introduction of steam ships brought to an end the age old sailing ships. Steam engine shipwrecks have been explored and excavated in many parts of the world. In India, preliminary explorations of two steam engine shipwrecks have been carried out in Minicoy waters of Lakshadweep and the remains have been documented (Rao et al., 1995-96; and Gaur et al., 1998). Similarly, Johnstone and Robinson (1993) have studied the remains of the *Indiana* steam engine shipwreck in Lake Superior; Nash (2002a) has studied the *Tasman* steamship (1873-1883) off the southeast coast of Tasmania. *SS Pomona* steam engine shipwreck has been studied by the Indiana University, USA (Personal communication: Charlie Beeker). In the '*Iron Ships and Steam shipwrecks*', McCarthy (1988) has provided insight into the exploration and excavation of iron and steam engine shipwrecks off Australia and other parts of the world. In the Hydrographic Chart No. 2020 a few shipwrecks are marked. Based on the information, survey was carried out off Amee shoals which led to the discovery of the shipwreck. This is the first steam engine shipwreck that has been observed in Goa waters. The shipwreck lies near the submerged rocks of 3 to 4 m high and the preliminary assessment shows that the ship was either approaching or disembarking from Marmagao port and might have run into the shoals and had been wrecked. The site is located close to Marmagao port channel and iron ore laden barges regularly ply over the area. In later period, some barges collided with the ill-fated wreck and wrecked over there. Except a cabin, hull and beams, the other remains of the barges have been salvaged. These remains lie further on the southern side of the boilers and highly corroded. The hull and beams of the barges are welded which makes easy to distinguish between steam shipwreck and barge remains. Steam engine shipwreck remains are mostly riveted. One of the possible causes for shipwrecks in shallow waters off Goa could be lack of knowledge of mariners' about the seabed topography of the region. Amee shoals wreck could have been caused by the grounding on the shallow submerged rocks due to human error in navigation (Sila Tripati et al., 2010a). From the preliminary exploration it appears that the wreck could be of a single ended Scotch boiler, steel hulled, triple expansion steam driven vessel. These types of drum boilers started from 1880s onwards and many companies were making these types of boilers. Based on the shape and size these boilers could be dated 1880 AD or of the mid 1900s. These three Scotch boilers indicate that the ill-fated ship was large in size, could be of 100 m in length (Personal Communication: John Riley). After the wreckage, a large portion of the hull might have been above the water for sometime, then crumbled and sank.

Typically, steam vessels are metal hulled and this could easily be determined from the site. It is presumed that Amee Shoals shipwreck could be steel hulled with wooden superstructure. These boilers are the last phase of development of fire tube boilers. Subsequently, these were replaced by diesel engines. The boilers were oil fired type not coal fired and bricks are used as insulators near the burners. The pipes are connected to the boilers to allow the oil to be pumped. It was not possible to quickly reduce the output of a coal fired boiler by shutting off the fires and before hand information was required to stop the engines but this was not the case with oil fired engines. In oil fired engines, fuel is stored in double bottom tanks. Steam heating coils were provided to make sure that oil will flow even in cold weather. Then oil is channelised to two vertical settling tanks and the presence of water in the oil is removed and drained off in one tank whereas the other tank is used to supply fuel to the furnaces with fuel pressure pump passing through filters, strainers and removes the foreign materials. The oil under pressure is passed through a heater in which steam raises to a suitable temperature for spraying and burning. Afterwards oil passes through a fine filter to the distribution box then to burners. By the 20th century, with technological advancement the use of higher pressures, more modern designs replaced the simple and old boilers. Subsequently, water tube boilers and diesel engines came into use. Usually firing conditions were different from ship to ship. In case of coal fired boilers, it was not possible to turnout the fires suddenly but cautioned before stopping engines. Fire was allowed to die back and by using damper, the draught was finally reduced. The ashes of the coal fired were wetted down with a hose and they were taken to ash pits. In oil fired boilers more care and cleanliness were crucial and heavy oil was used.

The single ended Scotch marine boiler consists of a cylindrical boiler shell with large diameter and short length, having two or more furnaces in corrugated fire tubes to withstand external pressure. Each furnace ends in a combustion chamber surrounded by water. The hot gases pass through a bank of fire tubes from the combustion chamber to the smoke box at the boiler front. Grate bars divide the furnaces into two parts: the upper part for the fire and gases and the lower part for the ash pit. The Scotch marine fire tube boilers contain a large quantity of water nearly six times more than normal water tube boilers. The water used to be kept to a level above the top of the tubes and combustion chamber which occupied all the space not taken up by the furnaces, combustion chamber, tubes, stay rods, stay bolts and steam space. Sometimes large Scotch boilers used to have as many as four furnaces. Like the Vertical Boiler the Scotch Boiler is a 'fire tube' design. In this case, a number of relatively small (9 cm diameter) metal tubes pass horizontally through the water cylinder and act flues. A boiler 3 m diameter and 7 m long would normally contain 137 individual horizontal tubes. These 'fire tubes' were arranged above the furnaces, but below the water surface (Bowden, 1971). Each furnace of Scotch boiler has a burner located in a register comprising swirler and shut off vanes to check the amount of heated air which comes from the forced draught installation. The oil is atomised by the burner and mixed with incoming heated air as well as swirls into the furnace as flame.

In the 1880s, most marine boilers were called 'Scotch Boilers', which were cylindrical, with the fire grate within a flue inside the actual pressure vessel. The next development (which was just starting for use in warships as Kipling was writing 'McAndrew's Hymn') was the water-tube boiler. In this, instead of the hot gases from the coal's combustion (or later oil) passing through a nest of small diameter tubes immersed in water, a nest of small tubes, filled with water, linking two drums, passed through the furnace. This arrangement was more efficient in transferring heat from the gases to the water and enabled the rate of steam generation to be more quickly altered. This was of value in a warship, where speeds might alter at short notice and by small amounts, to keep station, one ship on another. Most obvious is the single ended Scotch type boiler, which was the most common type, used for marine services in the late 19th century. By the 20th century, with the use of higher pressures, more modern designs replaced this simple and effective boiler. As boilers are fairly solid and quite heavy, they can usually be identified in the shipwreck sites which provide clues to the general orientation of shipwreck. Like boilers the engine usually also give an indication of the orientation of the vessel. Depending on the size of the vessel number of boilers may range more than one to dozens in a vessel. The shipwreck which is under discussion has three boilers which are usually sufficient to power the engine. In small vessels with a single engine the stoke-holes face towards the engine, allowing the engineer to act also as the fireman. Boiler fire bricks are commonly found in the steam engine shipwrecks and stamps on the bricks indicate the name of the company. The boiler bricks of Amee Shoals shipwreck are stamped 'FURNA......B84', which appears to be the name of the company believed to be British company. Boiler bricks of the SS '*Pomona*' shipwreck have been collected and those were stamped 'Carnegie'. The study of the bricks could lead to knowledge of the details of the famous brick manufacturing company in San Francisco. Carnegie manufactured bricks until it was burned down in the great 1906 fire (Personal communication: Charlie Beeker). It has been suggested by local people that during the Liberation of Goa three ships were engaged in the battle and those were sunk opposite Marmagao harbour. Probably this could be one among them. If, so what happened to the arms and ammunition? What about other two wrecks or this ship was also not involved in the war but was destroyed during the exchange of fire? Unless some clues are found, this cannot be confirmed. The shipwreck could not be a warship because warships of that period did not use Scotch Boilers. Scotch boilers take too long to raise steam for naval action (Personal Communication: John Riley). Warships used to have water tube boilers and the boilers were not fired by coal. If it had been a warship some kind of arms and ammunition could be expected in the wreck during exploration. It appears that the wrecked ship could be a cargo steamship belonging to the early 1900's and could be a British ship and the boilers oil fired the bricks used inside the boilers to direct the burning oil. The boiler bricks of the British warships used to have a broad arrow on them. During the Portuguese regime in Goa, Portuguese had leased the job of laying the railway line from Marmagao to Castle Rock in 1887 to a British Company (Gune, 1979). In order to construct the railway line the British transported steel from Sheffield, UK in several ships, hence it could be a British cargo ship.

The exploration has so far not yielded any ceramics, pottery and cargo remains. Except the boilers, water tank and other engine parts, no other remains have been found. The engine parts and boilers are well preserved. Its engine could be turbine, triple or quadruple expansion or a combination. The hull frames of the ship are corroded. All around the site copper pipes, tubes and wires are found. Due to non-availability of dateable finds it is difficult to exactly date the wreck and to know the provenance of the ship. It is felt that the cargo of the ship might have been salvaged soon after the wreck. Generally the name of the ship is inscribed on the bell which has not been found, which could have provided some evidence of the name and nationality of the ship. Similarly, the destination and other details of the ship are yet to be traced. Other details of the ship such as its length, width, masts, decks, anchor and her history are not known. Steam engine shipwrecks of era 1840-1920s are extremely fascinating to maritime archaeologists because of their historical importance. More number of steam shipwrecks has always been the target for salvage operators due to their highly valuable non-ferrous fittings and much has been lost. However, thousands of well-preserved steam engine shipwrecks are lying in deep water out of the reach of salvagers, but the future of shallow water shipwrecks is bleak.

Conclusion

The preliminary exploration of Amee Shoals brought to light encouraging results with the discovery of steam engine shipwreck remains. It appears that the ship could be a single screwed propeller driver, steel hulled, triple expansion steam driven vessel. Although identification and information of other details of the ship have not been made at present, further exploration may provide important clues. The wreck has a potential threat from the local divers who salvage the brass, copper and other metals from the site. The engine has lots of brass which is one of the first items to be salvaged, as the propeller. Till date not many shipwrecks have been explored in Goa waters, this is the first steam engine shipwreck; hence the exploration of this shipwreck is important in reconstructing the maritime contact between Goa and other countries. On the basis of boilers, the history of the ship may be reconstructed. During preliminary exploration boiler bricks and flanges have been collected from the site, further study of bricks may shed more light on it. The vessel probably was a cargo ship. Interestingly, no belongings of the crew, cargo and artefacts have been found, even a single sherd of pottery has not been recovered from the site. The region between Amee Shoals and Aguada bay has many potential sites for shipwreck exploration off Goa because of the presence of submerged reefs and shoals. The low lying rocky sub-bottom appears to be main cause of shipwrecks. Moreover, during the Portuguese regime, ships crossed these reefs and shoals to reach Old Goa, the main port of Goa. In India, very few shipwrecks have been surveyed to date but the rate of salvage is much higher. Unless this is curbed by the authorities nothing will remain for maritime archaeologists to undertake detailed studies of these wrecks.

6

Shipwreck Exploration off Sail Rock and Galgibaga

Archaeological and historical evidence indicates that over a period of time several developments have been noticed in shipbuilding technology i.e. sail to steam and wood to steel. In the course of maritime trade several ships have been wrecked due to numerous causes along the Indian coast which are broadly classified into two categories viz. manmade and natural. The coast of Goa is evidently dotted with many shipwrecks and most of them would be wooden hulled sailing crafts. Information on Sail Rock and Galgibaga shipwrecks was obtained from local divers and fishermen.

Description of the survey area and underwater exploration

The shipwreck of Sail Rock is marked in the Hydrographic Chart No: 2020 (Fig. 42) and the wreck lies near the reef. The reef gets submerged during high tide. The exploration area lies on the southern side of Grande Island and on the north-eastern side of Sail Rock. The shipwreck site separates Grande Island from Sail Rock (Fig. 43). The reef near the wreck site has a wider underwater extension. Bathymetric contours of Sail Rock indicate that the area is considerably deep. The seabed area is covered with sandstone blocks of various sizes which are scattered over a larger area. Overgrowth of soft corals and gorgonian was noticed on the shipwreck site. The seabed is full of silt and a thin layer of suspended material can be seen on the sandstone blocks. The water depth in Sail Rock varies from place to place, however the maximum water depth in the shipwreck region is 14 m. The position of the shipwrecks shown in the Hydrographic chart is marked in the GPS and navigated to the

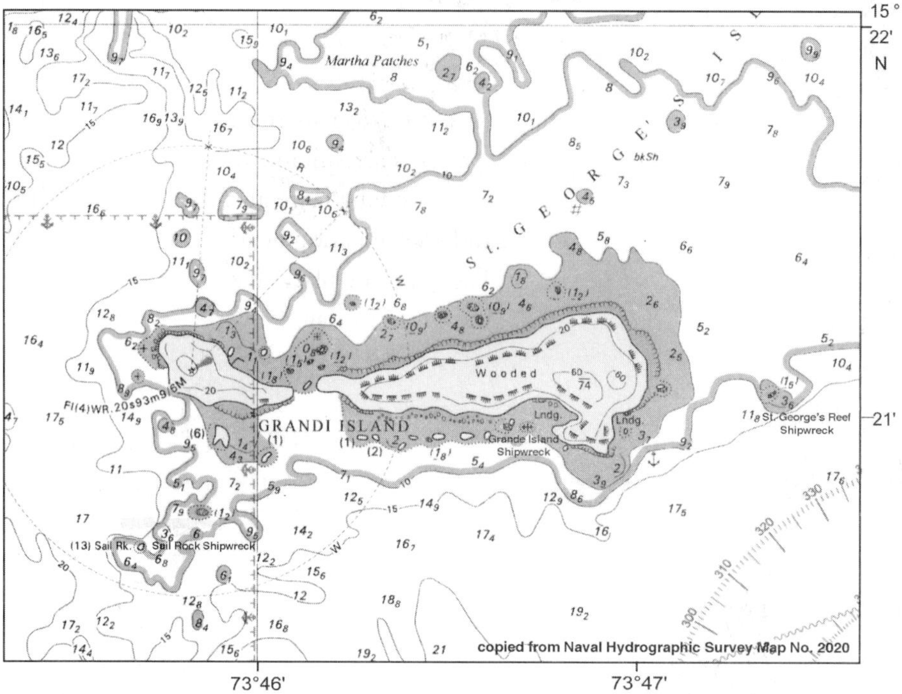

Fig. 42: Figure showing Sail Rock region off Goa

Fig. 43: Location of Sail Rock shipwreck

site and a marker buoy has been dropped at the wreck site on the basis of the GPS position. After locating the shipwreck, scuba diving was initiated in Sail Rock. In addition to the underwater exploration, natural features such as outcrop rock, seabed topography, water depth, movement of sediments and current direction in the region were noted during exploration. Swimline and circular search patterns are suitable for shipwreck survey; hence these search patterns are followed. Each dive covered a large area of about 5 to 7 m radius and attention was paid to documentation and identification of artefacts which would enable us to know the period of the ship and her flag. In some places hand fanning was carried out to expose the buried artefacts and parts of the ship in the sediment. In order to record the exact positions where underwater explorations are carried out and the shipwreck area, co-ordinates were taken by Magellan model NAV 5000 DLX GPS.

Shipwreck Finds

The shipwreck remains off Sail Rock are scattered over a larger area. In the course of exploration iron anchor, propeller, rudder, engine parts, mast, cabins, winches, tile piece, and buckle, were found and these were photographed. In addition to these finds, an iron shot was also noticed at the site. The major portion of the shipwreck remains lies on the western side of the site. The front portion of the ill-fated ship is totally damaged whereas the rear portion is intact.

Iron Anchor

A stockless iron anchor was found between the rudder shafts and mast. Only the head and shank of the anchor are visible. Remaining portion of the anchor is covered with iron frames of the ship and buried in the sediment hence it is difficult to notice any marks and symbols on it. The anchor is lying in the South - North direction i.e. the head is on the South and the shank is on the North side. Whether the iron chain was attached to the shank or not, is difficult to ascertain. Moreover, it is also difficult to say whether the ship was carrying only one anchor or more.

Propeller

The four blade propeller (petals) of the ill-fated ship was found, among which three blades are visible and the fourth one is either broken or buried in the seabed. The length of each propeller blade is approximately 2 m and the base width of the propeller is 1 m (Fig. 44). All these blades are damaged by the local divers and cut marks are visible on each of them. As the propeller is made of non-ferrous metals, marine growth is not noticed on it. A piece of metal has been cut off from the eastern side of the propeller blade by some local divers.

Fig. 44: Propeller of shipwreck off Sail Rock

The gap between propeller hub and shaft bearing is clearly visible. The diameter of the propeller shaft is 140 cm. A flange has been provided to the shaft with five rabbets which are clearly visible; the distance between the rabbets is 20 cm and marine growth is present on them. There is a gap between the propeller and body of the ship. The rudder is intact and connected with the propeller. The shaft on which the propeller is attached has some equal distance holes of which only three holes are visible.

Propeller shaft

Five pieces of propeller shaft were noticed on the wreck site, all made of hardened steel and the spline was noticed on the end of the shaft (Fig. 45). This spline allows the propeller to

Fig. 45: Propeller shaft of shipwreck off Sail Rock

be mounted on the propeller shaft without slipping free. Each is 6. 70 m long and the circumference is 1.10 m. Each shaft has couplings which have 6 holes of 7 cm width. The diameter and thickness of the coupling is 70 and 10 cm. Marine growth is present on them. Two pieces of shaft are connected to the propeller whereas the other three pieces are separated and lying together. *Propeller* shaft is attached to transfer the power from the engine to the propeller. It then runs through the hull and into a bearing just ahead of the propeller. The shaft must run true and straight, free of any bends or it will vibrate the vessel as well as pre-maturely wear out the bearings and seals. The propeller shaft runs from the engine through a seal in an inboard engine application.

Mast

A 12.65 m long mast of the ship was found which is lying 10 to 12 m on the western or NW side of the shipwreck site. The mast has been separated from the shipwreck. Its circumference is 2.6 m. Some portion of the mast is broken probably after hitting against rocks and the remaining portion is undamaged. The 6.7 m long ladder attached to the mast is also intact. It appears that the ship hit with such force that its mast was dislodged from the ship and lies 10-12 m away from the ship. Some (U) shaped beckets are noticed on the mast and marine growth is also present on the mast.

Cannon ball, Buckle and Flooring Tile

In addition to these finds a cast iron cannon ball weighing 9.2 kg found below the cabin (Fig. 46). Its surface is rough and uneven. No marks and fuse holes are found on it. The cannon ball has been documented for further study. A rectangular (4 x 3 cm) shaped copper buckle was found near the mast. The buckle has clips for holding the leather. On its left hand corner there is a letter 'D' enclosed in a triangular identifying mark (Fig. 47). The buckle is twisted and marine growth is present on it. No other remains were found along with the buckle. During exploration a rolled broken piece (9 x 7 x 1.5 cm) of a flooring tile was found close to the propeller shaft. On

Fig. 46: Cannon ball found during exploration of shipwreck off Sail Rock

Fig. 47: Copper buckle found near the mast of Sail Rock shipwreck

its upper side prominent diagonal lines are present with regular intervals whereas the lower side has straight lines with round depressions which are probably meant to provide better grip (Fig. 48). Marine growth is present on it. Probably these tiles were used on the floors of common bath, wash room or kitchens.

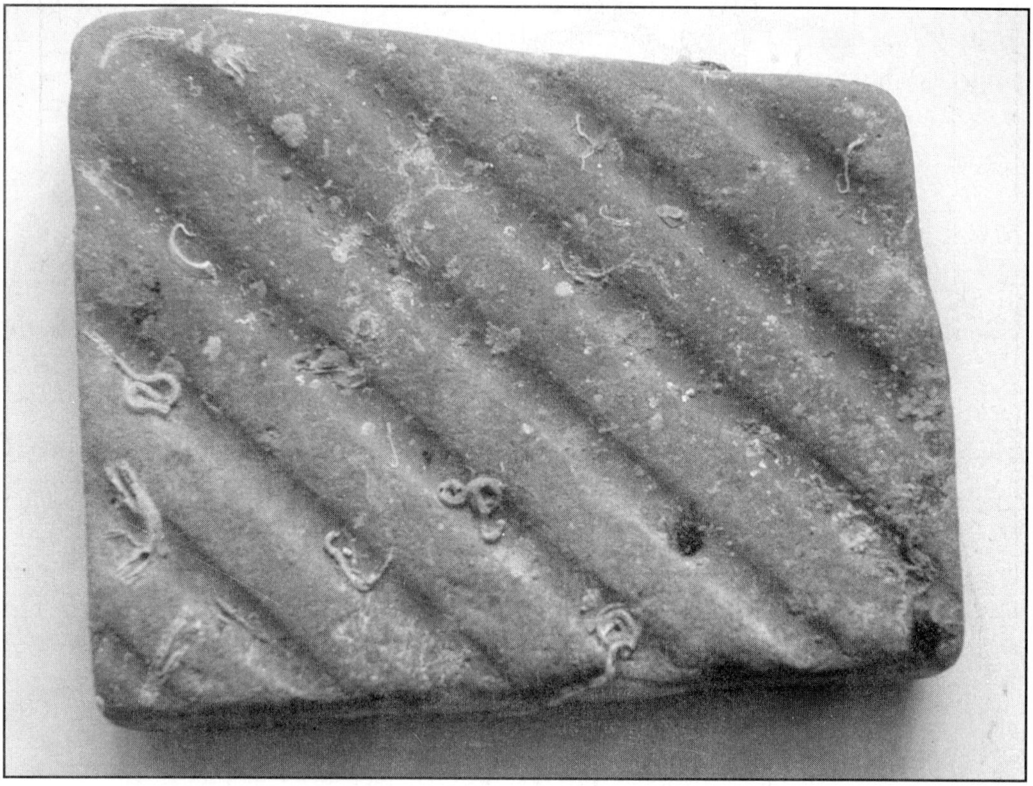

Fig. 48: Broken piece of tile found near the shaft of Sail Rock shipwreck

The Galgibaga shipwreck

The Galgibaga shipwreck which killed 10 of the 23 Chinese sailors is more than 60 years old. The engine of the ill-fated ship failed in the mid sea off Galgibaga while carrying ore and sailing from Marmagao port to Hong Kong. Local people believe that the Chinese ill-fated ship ran aground during monsoons off Galgibaga beach sometime in 1950 (Fig. 49). The ship sank away from the shore in deep seas, and then drifted nearby to Galgibaga coast. Villagers say that all the dead sailors were buried behind and outside the present Galgibaga cemetery. While five sailors reportedly left Galgibaga via land to reach their native destination, about eight sailors stayed back for six more years in the village with the hope of retrieving and making the ship navigational again. But this has never happened

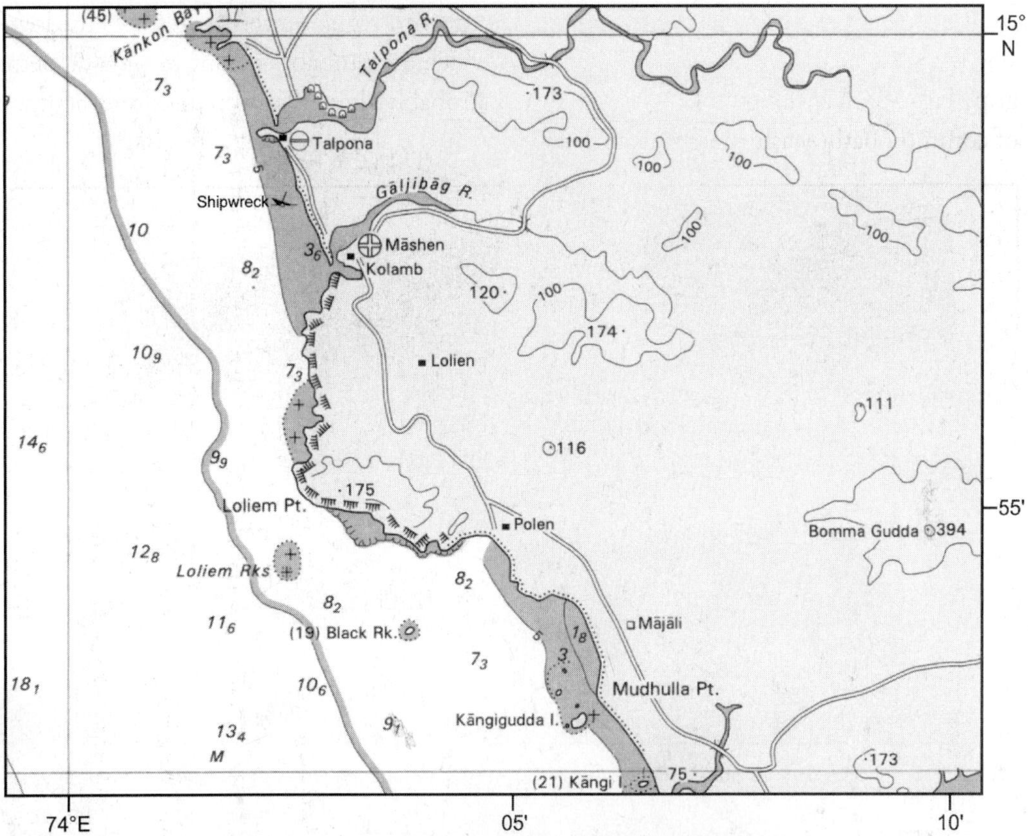

Fig. 49: Figure showing the location of Galgibaga shipwreck

and all the Chinese sailors left Galgibaga as their efforts proved futile. Sources also informed that besides ore, the ship also carried a stock of its booty and other costly items were willingly allowed to be taken away or distributed by sailors to their benefactors from the village. After all the sailors left some miscreants as well as local divers later decamped with materials which were available in large quantity inside the ship, including the pipeline and other costly gadgets from the ship. It has been observed that the shipwreck remains have been scattered over a large area. Preliminary explorations carried out nearer to the shipwreck to record the remains. Due to bad weather and poor visibility exploration could not be carried out further. However, the remains such as iron frames of the shipwreck can be seen during low tide at many places (Fig. 50).

Discussion

The Sail Rock shipwreck might have occurred due to lack of knowledge of seabed topography of Goa or the ship might have drifted, hit against the submerged rock and wrecked

Fig. 50: Remains of Galgibaga shipwreck can be seen during low tide

subsequently. The main causes of shipwrecks in Goa waters are because of outcrop of laterite reefs and shoals, storms, lack of knowledge of mariners on the seabed topography of Goa, etc. The exact cause of the loss of the ship is still unknown. Major portion of the bow part is missing whereas rear portion of the ship is in good condition. A number of stone boulders of various sizes were noticed in the adjoining area of the shipwreck. The shipwreck is ransacked by local divers. All the non-ferrous metal, engine parts and other parts which can fetch some money have been taken by them. Marine growth is present on the remains. Major portion of the shipwreck remains are lying on the western side. During exploration neither datable artefacts such as ceramics, pottery nor cargo remains of the ship were found. However, from the size of the propeller it could be suggested that the shipwreck off Sail Rock could be late 19th or early 20th century. The broken terracotta tile is probably from the galley. It could be a cargo ship whose cargo might have been washed out or salvaged soon after the wreck. The hull frames of the ship are highly corroded. A number of iron objects are lying scattered over the region. The name of the ship is inscribed on the bell which has not been found, the bell could have provided some evidence on the name of the ship. Similarly, the destination, period and other details of the ship are yet to be traced.

7

Stone and Iron Anchors from Goa Waters

Anchor is one of the earliest tools of navigation and developed by men to secure their craft. In early times crafts were secured by dragging on the lake or seaside as far as possible. When they understood it is necessary to hold the craft stationary for long time either due to catch, tides, currents, etc. then they needed some means of holding the craft, without being wrecked and anchor came into use. Anchors provide critical clues on the ancient maritime activities of the country. Stone anchors are as old as shipping. With the beginning of shipping either in lakes, rivers and seas mariners used according to their requirements various shapes and sizes of stones as anchors. Further, anchors are the representative of the existence of ancient harbour, shipwrecks, anchorage points and trade routes. The study of stone anchors shows that initially stones of required shapes and sizes were used by mariners as anchors. For instance a required size of stone (with and without holes on rock) was tied with rope and used as an anchor (Fig. 51), these were

Fig. 51: Dead weight stone anchor

the primitive ones and still used in different parts of India, then hook shaped wooden shanks (Fig. 52) tied with a suitable sized long stone which are known as Killick anchor. According to the Dictionary of Newfoundland English, 'a Killick is an anchor (Fig. 53) made of an elongated stone encased in pliable sticks bound at the top and used for small crafts. In other words, four to six wooden sticks were tied together and other ends of the sticks were inserted into the

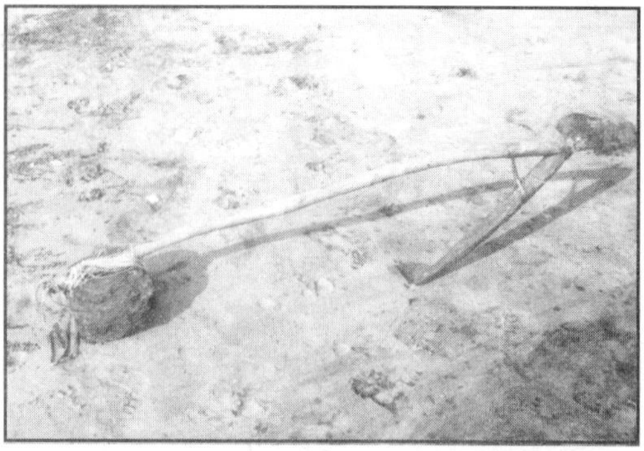

Fig. 52: Hook-shaped wooden shanked stone anchor

Fig. 53: Killick anchor

crosspiece of hard wood which was provided at the base. Suitable sized stones were kept in the crosspiece. The points of the crosspiece secure to the seabed. It has been suggested that in the River Ganges Killick anchors were used. Over the period of time stone anchors of various types were invented by the mariners as required for particular type of seabed. For example, dead weight stone anchors such as single hole types came into use; subsequently composite and Indo-Arabian types were developed and flukes were introduced, then ring stone (mushroom) type of anchors simultaneously were brought in (Fig. 54). Composite

Fig. 54: Different types of stone anchors recovered from Indian waters

anchors appeared to be suitable for sandy bottoms whereas Indo-Arabian anchors were efficient for rocky and coral bottoms. Bigger and heavier stone anchors introduced and used once the size, carrying capacity of ships increased and long distance maritime trade came into existence. Later on stone anchors were replaced by lead stock and further by iron

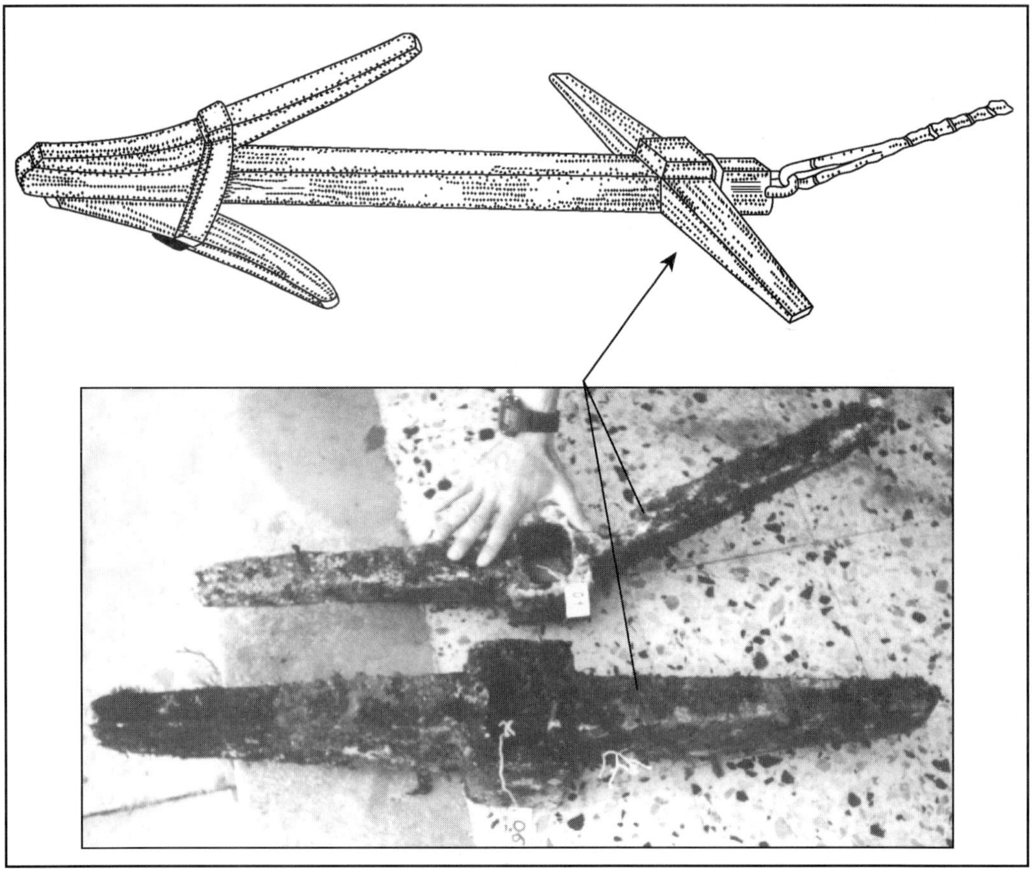

Fig. 55: Lead stock anchor

anchors. A number of lead stock anchors (Fig. 55) with and without wooden stock have been recovered from the Mediterranean Sea and other parts of the world. However, so far no lead stock anchors have been reported from Indian waters, whereas iron anchors were introduced in Indian waters in the 16th and 17th centuries by the Europeans (Qaisar, 1982). Along with iron anchors various types of stone anchors were in use in Indian waters for centuries.

A great deal of new evidence concerning with maritime archaeology of India has come to light in the recent years from onshore and offshore explorations and excavations. Amongst other finds stone and iron anchors are the keys to understand the history of maritime trade and commerce, overseas contacts and shipping activities of the region. Maritime archaeological explorations along the coast of India have brought to light varieties of stone anchors from Goa (Sila Tripati et al., 2003a), Gujarat (Gaur et al., 2002; and Gaur et al., 2007b; and Sundaresh et al., 1999; and Sundaresh et al., 2002), Kerala (Sila Tripati et al., 2005b), Lakshadweep (Kapitan, 1994; and Sila Tripati, 2009) and Maharashtra (Sila Tripati and Gaur, 1997; and Sila Tripati et al., 1998a; and Gaur et al., 2007a) on the west coast and Tamil Nadu (Athiyaman, 2001) and Orissa (Sila Tripati and Patnaik, 2008) on the east coast of India. Ever since underwater explorations were carried out in Sunchi Reef, Grande Island, Baga and Aguada waters (Fig. 56) varieties of anchors namely Indo-Arabian,

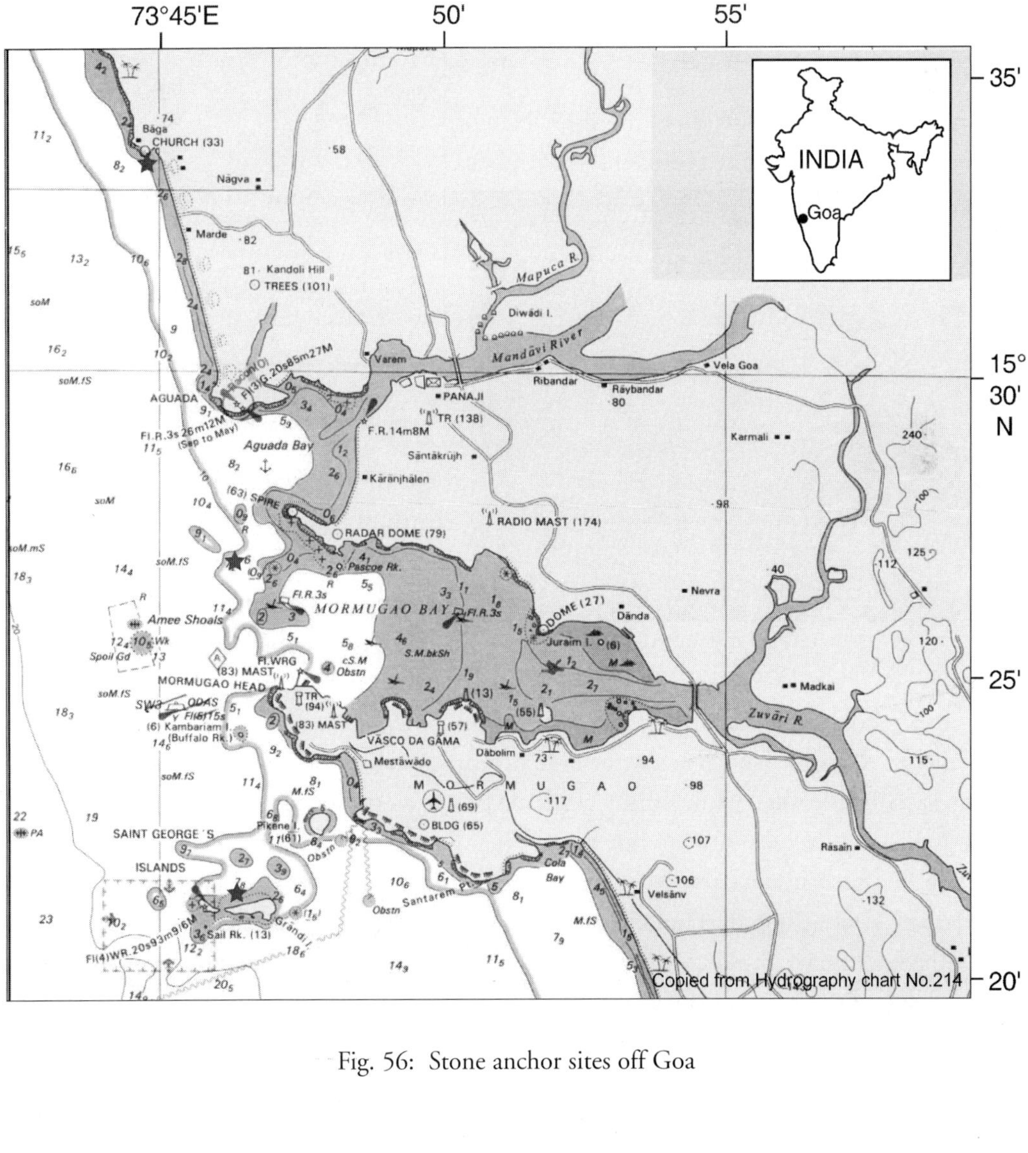

Fig. 56: Stone anchor sites off Goa

ring stone anchor, a pyramidal type (appears like a broken Indo-Arabian type of stone anchor) (Fig. 57) and iron anchors have been recovered. In addition to these iron anchors, four iron anchors have been recovered while dredging the sandbar of Aguada bay. Similarly, two more iron anchors have been housed in the Archaeological Museum, Old Goa, which resembles with Sunchi Reef and Aguada waters finds.

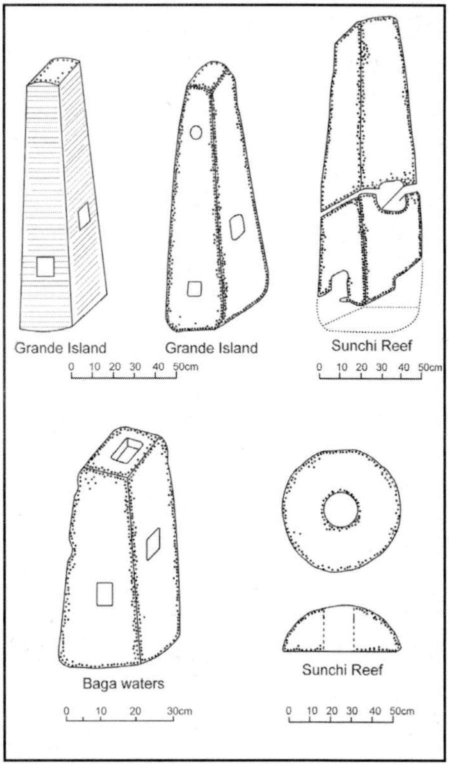

Fig. 57: Stone anchors recovered from Goa waters

Indo-Arabian type of stone anchor from Sunchi Reef

Exploration on the eastern side of the shipwreck of Sunchi Reef brought to light a broken Indo-Arabian type of stone anchor from 5 m water depth (Fig. 58) which was trapped in between the lateritic outcrop. The anchor has a distinct crack on its upper side. Its surface is uneven and marine growth is present on it. It has only one lower square hole, which is highly abraded and the other square lower hole is broken. The stone anchor does not have the hawser hole meant for cable and it is in a fragile state. The anchor is so brittle that while lifting again it broke at the other lower hole. Its present length is 114 cm, weighs 100 kg, but the original length was considerably longer than it is now, due to which it has been reconstructed to arrive at its original shape and dimension (Sila Tripati et al., 2003a). The anchor is made of Amygdaloidal basalt. The amygdules within the basalt have presence of secondary mineralisation of calcite and shows twinkling optical characters of two sets of cleavages (Sila Tripati et al., 2010b). The rock of the anchor could have been sourced from Amygdaloidal basalts of Deccan Trap age capping the Jurassic limestones in Kachchh, Gujarat (Merh, 1995).

Fig. 58: Indo-Arabian type of stone anchor from Sunchi Reef (Scale in 5 cm divisions)

Ring stone anchor from Sunchi Reef

A ring stone (mushroom) anchor was found lying on laterite boulders on the western side of the guns of the shipwreck off Sunchi Reef. Its surface is uneven and overgrown with oysters and shells (Fig. 59). A distinct crack line is visible on this anchor. The crack might have occurred after hitting the rocks of the seabed. A big chip has come out at one place of the anchor, this must have occurred while making the anchor. In some places chisel marks are visible on the anchor. The anchor weighing 76.6 kg was found in 5 m water depth (Sila

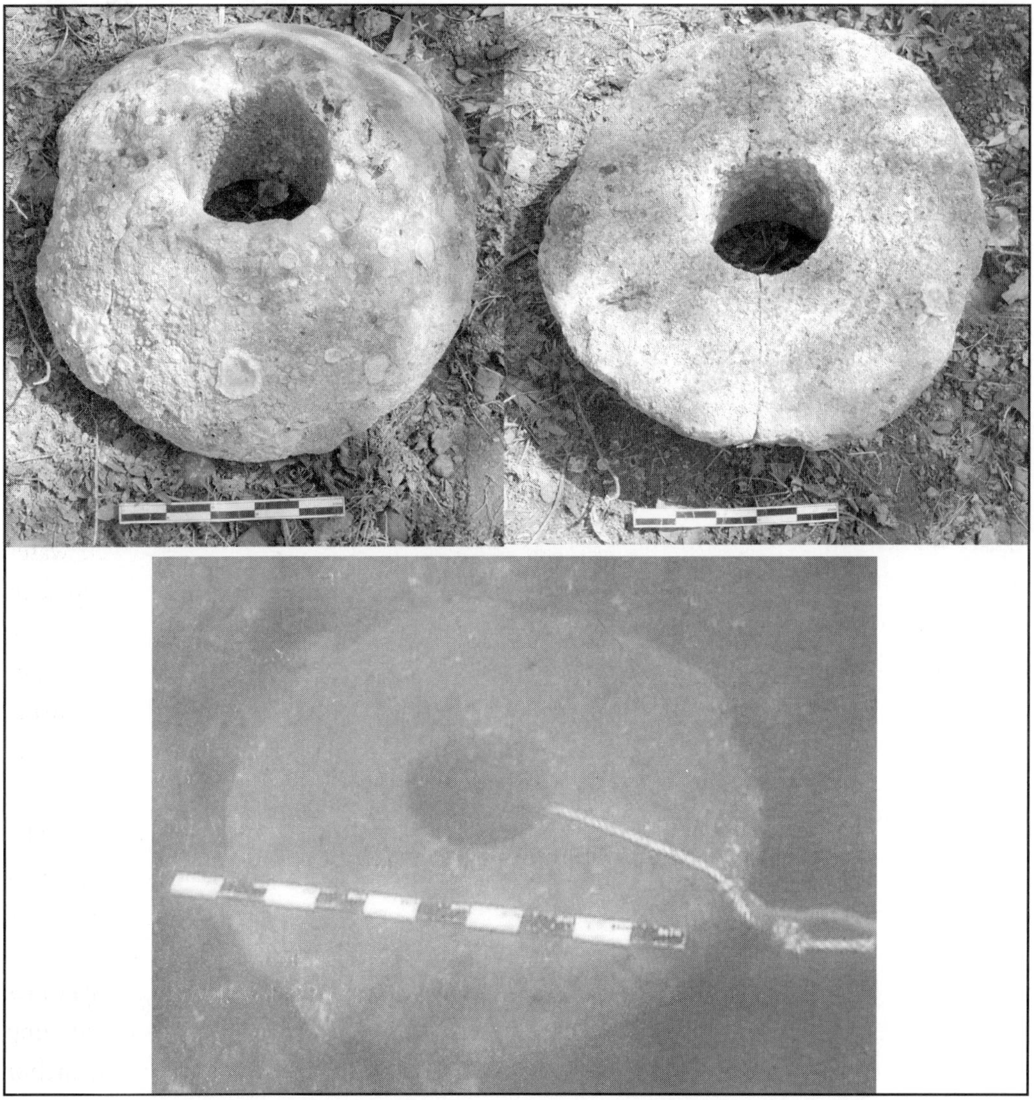

Fig. 59: Ring stone anchor from Sunchi Reef

Tripati et al., 2006b). This anchor resembles other ring stone anchors recovered from Dwarka, Bet Dwarka and Somnath (Gaur et al., 2002) and recently also reported from Miyani and Visawada waters of Gujarat. This anchor is made of Actinolite schist. The actinolite schist is composed of green coloured pleochroic actinolite laths arranged in parallel fashion along with dark black coloured spinels, which endowed the schistosity to the rock. The source material of the anchor is plausibly derived from north-eastern Gujarat bordering Rajasthan and reported as Rakhabdev Ultramafic suite of metamorphosed serpentinites (Bhu et al., 2006; and Merh, 1995). Similarly, in order to understand the provenances Iyer et al (2006) analysed composite, ring stone and two holed stone anchors off Dwarka coast and the study evidenced that composite and two holed anchors are made of locally available limestone and calcareous sandstone respectively, while the ring stone anchor is made of basalt, which could have been brought from nearby coastal dyke either from Junagarh or Girnar hills.

Indo-Arabian stone anchors off Grande Island

Two Indo-Arabian types of stone anchors have been retrieved off Grande Island. This anchor (Fig. 60) was found in 1997 approximately 300 m away from the jetty of Grande Island in 7 m water depth. Deep and distinct chisel marks can be seen on the surface of the anchor. The upper side of the anchor is round instead of square and it tapers upwards after the upper square hole. In comparison with size of the anchor the size of the holes are smaller and it weighs 113 kg. Coral growth is present on one side between lower square hole and upper square hole, besides other marine growth is also noticed on its surface. It is made of Granodiorite (Sila Tripati et al., 2010b). The granodiorite, a hypidiomorphic holo-crystalline

Fig. 60: Indo-Arabian type of stone anchor recovered from Grande Island (Scale in 5 cm divisions)

rock, is composed of quartz and altered feldspars and rich in alkali metals with Na_2O and K_2O together constituting ~7% with high SiO_2 content (~70%). Alkali-rich igneous rocks are reported from Murud-Janjira, along the Maharashtra coast (Desai and Bodas, 1984) and Phenai Mata hill in Gujarat (Merh, 1995) which might be the source for the anchor. The other Indo-Arabian stone anchor (Fig. 61) was found in 2001 in 4 m water depth around 1.9 km north of the earlier anchor. This anchor does not have the upper hole; however its two lower holes are present and it weighs 130 kg. The anchor has been neatly trimmed by chisel leaving parallel lines running all around the anchor. However the holes are chiselled smoothly without parallel lines. The anchor tapers upward although it is square in section and some portion of the apex side is chafed. A big piece has been chipped out on one side of the base, whereas other three sides are uniform. Marine growth is noticed on this anchor. Both the anchors are different in shape, size and nature of chiselling shows they were not used extensively and therefore their edges are sharp (Sila Tripati et al., 2003a). The raw material of the anchor is Quartz-biotite schist with quartz and biotite as main mineral phases along with minor amount of feldspars. Accessory minerals like tourmaline, rutile and zircon are observed within the rock. These rocks were reported near Karwar in northern Karnataka coast of India (Radhakrishna and Vaidyanadhan, 1994).

Fig. 61: Indo-Arabian type of stone anchor from Grande Island (Scale in 5 cm divisions)

Stone anchor off Baga

The underwater exploration off Baga in 3 to 6 m water depths led to the recovery of a pyramidal type of stone anchor. At first glance it looked like a broken Indo-Arabian type of stone anchor. The anchor has two rectangular holes at lower side and a rectangular cutting on the apex (Fig. 62). The total length of the stone anchor is 53 cm, weighs 51 kg. Its

Fig. 62: Stone anchor found during underwater exploration off Baga (Scale in 5 cm divisions)

broader side is trapezoidal in shape and narrower side is nearly rectangular. There are some worn out marks on the anchor surface. The apex cutting of this stone anchor is uncommon and it must have been provided to complete the stone anchor. Hard wood or molten lead must have been inserted in the apex cutting to fix a metal hook. Sometimes for easy retrieval cords were passed through to fasten both the protruding parts of the wooden bars at the sides and knotted together (Sila Tripati, 2002b). The result of this adoption is new and perhaps it may be the occasional production. There is no connection between the rectangular cutting at the apex and the upper hole, which is generally noticed on other pyramidal anchor stones (Frost, 1989; and Tzalas, 1999). The anchor rock is composed of actinolite and is much richer in spinels. The rock has also high content of loss on ignition (LOI) (~11%) due to actinolite dehydration (Sila Tripati et al., 2010b). The composition is similar to the rocks reported from the Rakhabdev Ultramafic suite of metamorphosed serpentinites (Bhu et al., 2006; and Merh, 1995).

Iron anchor off Sunchi Reef

The iron anchor with 3.10 m long shank having two flukes was noticed in between the outcrop of lateritic boulders on the northeastern side of the shipwreck. The length of each fluke is 1.10 m. One fluke is exposed whereas the other one is inside the crevices of the laterite hence it is very difficult to document the anchor. The shank has a hole meant for metal ring in which cable was tied, however the ring is missing. The farthest end of the shank is square in shape in which the wooden stock was provided, no remains of the wooden stock was found in the nearby area. The wooden stock might have disintegrated over the period of time because the shank is lying on the laterite boulder and the region falls under the high energy zone. A thick layer of marine growth is noticed on the anchor. This anchor resembles other iron anchors in shape and size which were collected from Aguada bay, Aguada bar and others displayed in Archaeological Museum, Old Goa.

Iron anchors off Aguada Bay

The local fishermen frequently state that their fishing nets often entangle at *'Satbahu'* region of Aguada bay where the water depth is ± 14 m (Fig. 63). 2. 65 m long iron anchor with 2. 53 m long nailed wooden stock was found during fishing off Aguada bay (Fig. 64). The state of the anchor is good and intact because it was buried in the seabed; however, marine growth is partly present on it. In order to joining the stock and keeping it firmly together, nine iron nails or pins have been used and the shank has been placed at the centre of the stock. Out of nine nails four have been used around the shank for keeping it firm and sturdy, the remaining five nails have been used on the other part of the stock. Among these five nails, four nails were noticed on the intact portion of the wooden stock and only one is used on the teredo affected side. Uniform distances have not been maintained between nails nor did iron bands provide further fastening the wooden stock which generally is found in similar type of anchors belonging to the later period. The anchor has also an iron ring meant for tying cable, which is highly corroded. This is the first iron anchor with a wooden stock of its kind ever reported from Indian waters (Sila Tripati et al., 2003a; and Sila Tripati et al., 2005a). One side of the wooden stock was buried in the seabed which is intact, whereas the exposed portion of the wooden stock is highly affected by woodborers. Thick yellowish cement like coat has been applied all around the surface of the anchor probably to protect it from marine environment.

Analysis of coating material of the anchor

The microscope observation of the coating material shows that it is buff colour, comprising fine sediment of mica and shell material. It also appears that magnesium is mixed with

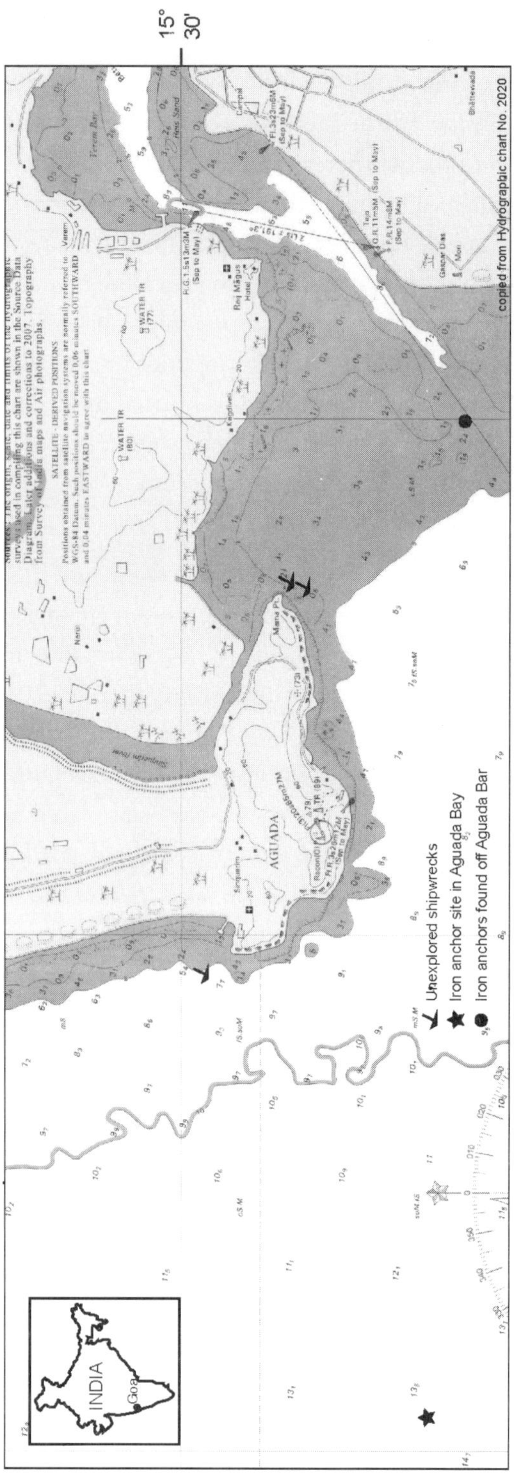

Fig. 63: Iron anchor sites off Goa

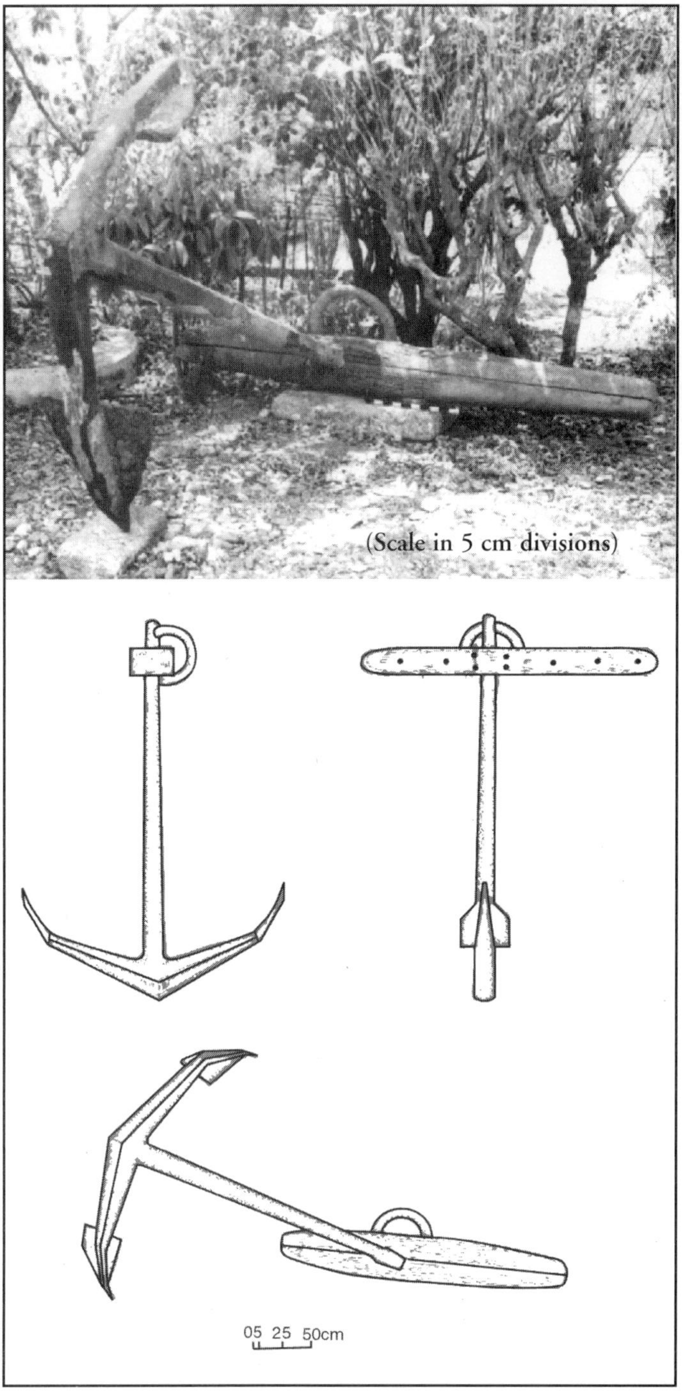

Fig. 64: Iron anchor with wooden stock from Aguada Bay (Scale in 5 cm divisions)

calcium, which becomes hard and this mixture can be used as a preservative and it does not dissolve in marine environment. The Infrared Spectophotometry analysis of the coating material carried out on Shimadzu FTIR model 8201PC shows the presence of some organic material, which could be a marine organism deposited on the stock during the later period. Since there is marine growth and a coat on the anchor hence no marks are visible on it. The anatomical analysis of wood indicates that it belongs to the *Tectona Grandis* species for which the trade name is teak and the Radiocarbon date of the wood suggests that it is 2500 ± 410 years BP; its calibrated age ranges 2950-1950 years BP (Sila Tripati et al., 2005a). One more Admiralty long shanked type of iron anchor was found in 11 m water depth opposite to the lighthouse of Aguada bay (Fig. 65) in January 2008. The seabed of the area consists of full of laterite boulders and fine silt mixed with clay. Subsequently, the anchor was retrieved for study and analysis. Attempts were made

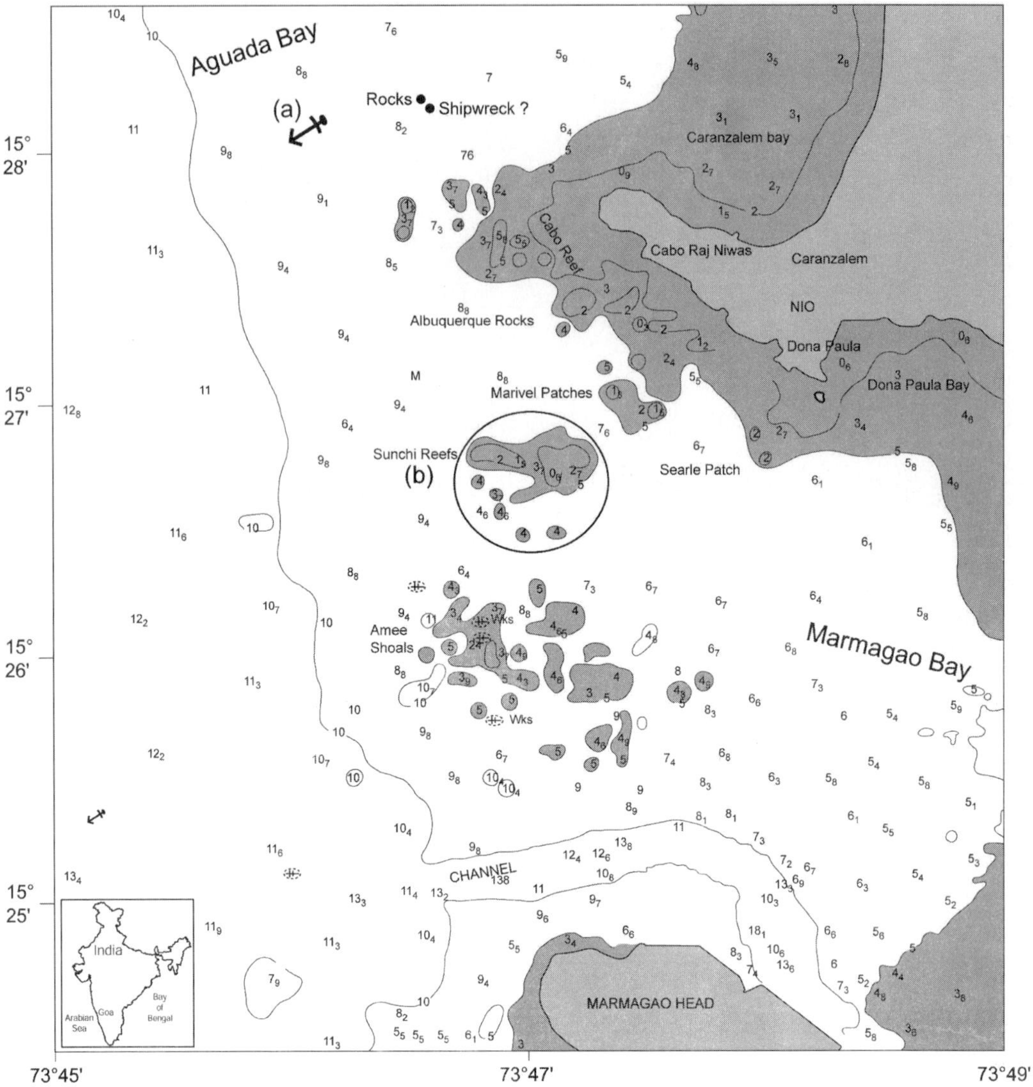

Fig. 65: Location of iron anchors recovered from Aguada Bay

for explorations in the area but due to poor visibility nothing could be recorded. The state of the anchor is good, its wooden stock and two flukes are intact, but its surface is uneven. On the surface of the anchor an application of a thick layer coat can be seen. However, the coating has been partially washed out on the wooden stock but it is present on the flukes, shank and nails of the wooden stock (Sila Tripati et al., 2009). The maximum length of the anchor is 3.30 m and each fluke measures 1. 35 m; the diameter of the shank varies from 45 to 47 cm. The maximum length of the wooden stock is 5. 37 m, made of two timbers joined together. The stock is much longer than the anchor and the thickness of the upper and lower timber also varies in 10 cm (Fig. 66). The upper timber is highly decomposed and deteriorated by the woodborers (teredo) and marine shells are embedded on the stock at different places. Wood pieces are crumbling out because of its decay condition. Since the retrieval of the anchor from the seabed it has been exposed directly to the sun, hence the timber has turned black like a burnt wood. The lower timber is bit harder, stronger and not disintegrated like the upper timber. However, its outer core is deteriorated. Nine iron nails (pins) have been used to join both the timbers of the stock and the shank is placed at the centre of the stock. All the nails are used at varied distance. Among the nine nails, five have been used on the right side and three on the left side and one is very close to the shank. All these nails are in good condition and covered with the coating layer. Even, in this anchor no iron bands are used for further fastening of the wooden stock. A 2.5 cm thick and 55 cm diameter iron ring has been provided in the anchor for tying cable or rope, but the ring has been corroded at certain places.

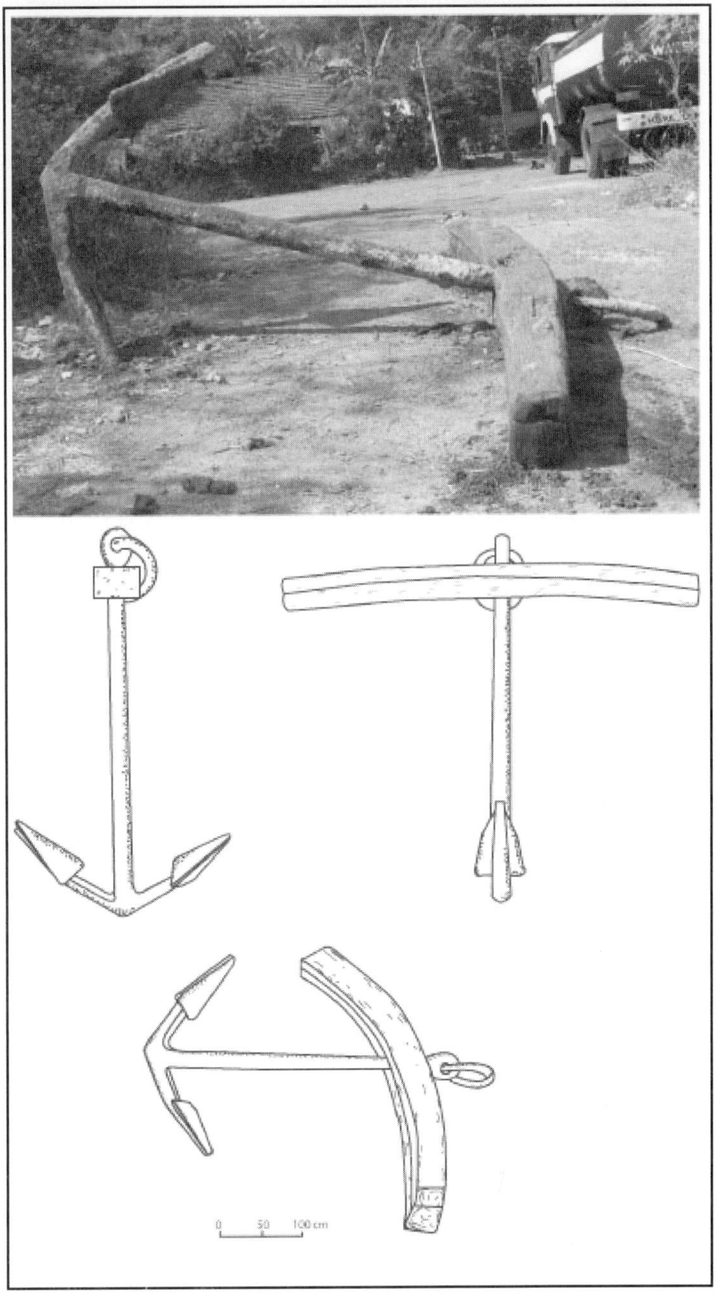

Fig. 66: Iron anchor with wooden stock from Aguada Bay

Analysis of coating material of the iron anchor

SEM and EDS analysis of the coating material shows that the composition of the coating layer is homogenous indicating the use of well sorted sands for protective coating on the anchor. The composition of coating material is analogous to beach sands rather than river sands. River sands are generally more variable and divergent in composition than the beach sands due to high energy waves swash on the beach front. Most beach sands, however, are composed of grains of the mineral quartz due to its higher relative density. In addition to the predominant quartz, beach and river sands contain other lithics such as feldspar, mica, hornblende, ilmenite, magnetite, garnet and also a variety of trace minerals such as iron/ titanium oxides, zircon, monazite and calcite. Substantial TiO_2 (0.81-2.07%) found in the iron anchor coat may be due to the presence of garnet as garnet is commonly present in lithics derived from hinterland. Titanium may be present in trivalent state in garnet along with Fe and Al. Based on the variations in the contents of oxides of Na, Mg, K, Ca, Ti and Fe in the coats of wooden stock and anchor, it appears that source material from two different beach sands may be used for these coatings (Sila Tripati et al., 2009).

Iron anchors off Aguada Bar

In addition to the above two iron anchors, the Archival records of the Goa State Archives, Panaji and India House in Lisbon mention that several ships have been wrecked off Aguada bar of Goa. The bar lies at the estuary of the River Mandovi which has been often referred in the Portuguese Hydrographic charts as *Barra de Agoada* and the water depth at the bar used to be less than 3 m. In 1977- 82, the Captain of Ports, Government of Goa, carried out dredging at this bar for smooth movement of all kind of vessels. In the course of dredging four iron anchors, many iron parts of ships and timber pieces of various sizes were retrieved from the site. Presently, these anchors and iron objects are housed in Goa State Museum, Panaji. Descriptions of all four anchors are mentioned below: The biggest iron anchor has 3.70 m long shank, without having wooden stock. It appears that the stock might have decomposed but the hawser ring is present. The prows are highly corroded (Fig. 67a). Iron anchor could be datable to the 16th-18th centuries. The other iron anchor has 2.70 m long shank with a hole at the end which was meant for the wooden stock; which is missing (Fig. 67b). Even, this anchor could be datable to the 16th-18th centuries. The state of preservation of the anchor is good however its prows are slightly corroded. Preservation coat is not visible on their surfaces. These anchors resemble each other and similar to the anchors found in Sunchi Reef and Aguada bay. The length of the shank of another anchor (Fig. 67c) is 2. 35 m and its ring is present but corroded. One prow of this anchor is missing. Instead wooden stock flexible iron stock is provided and the length of the stock and shank is same. The length of the shank of the fourth iron anchor is 2.50 m (Fig. 67d). The anchor stock is missing. These two anchors (Fig. 67c-d) are similar in shape and datable to the 18th century. None of these anchors have any weight or manufacturing marks.

Iron anchors displayed at Archaeological Museum, Old Goa

Two iron anchors have been displayed in the Archaeological Museum, (ASI), Old Goa. Initially these anchors were housed at the Institute of Menezes Braganza Hall, Panaji, then taken to the St. Cajetan Convent, Old Goa, subsequently shifted to Bom Jesus Church and at last housed in the Archaeological Museum, Old Goa. Both the anchors are similar in shape and size and datable to the 17th-18th centuries, but their origin, finding place and other associated information are enigmatic. The length of the one anchor shank is 3. 35 m and a square iron band has been provided on the shank. The prows and small portion of the anchor shank is corroded (Fig. 68). The ring attached to the shank for tying cable is present. The provision of the notch near the ring is to fix the stock. The other anchor is smaller than the previous one. Its shank is square in shape and length is 3.15 m and iron

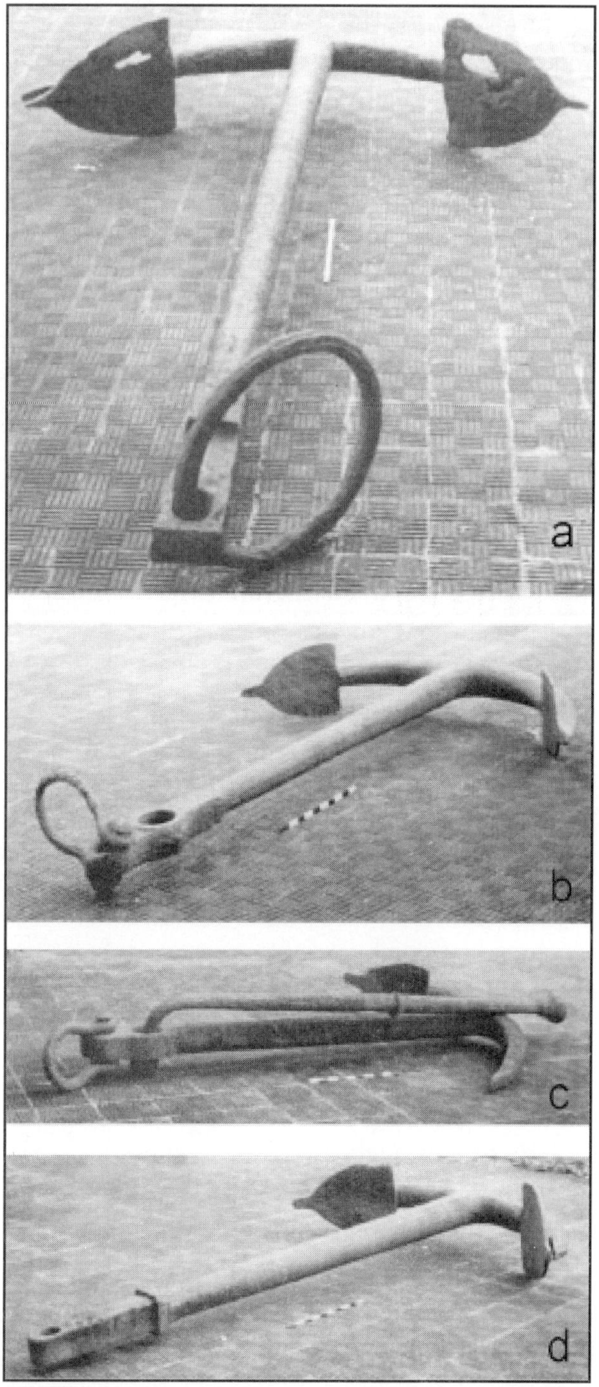

Fig. 67: (a-d) Iron anchors found during dredging in Aguada Bar (Scale in 5 cm divisions)

ring meant for fastening hawser is missing. About 1 m of the shank is worn out, one prow is missing and other one is highly corroded. Even, in this anchor the notch is present for fixing the stock. Wooden stocks of both the anchors are missing. Preservative coating has not been applied on both the anchors.

Discussion

The evolution and development of anchors were closely related to maritime history of the country. The study of each type of anchor (from stone to iron) shows that they have special character, origin, development and suitable for particular type of seabed. Over the period of time the size of ships have changed, carrying capacity of ships has increased and nature of maritime trade has altered. Under this scenario the shape and size of anchors and raw materials have also changed from stone to lead then to iron. In the beginning, dead weight and Kellick anchors were used followed by the hook and movable stock anchors (Puech, 1984). Then two hole and composite stone anchors were introduced. Later on Indo-Arabian and ring stone anchors were brought in. The common feature of Indo-Arabian type of stone anchor is that these are oblong in shape, tapers upward, having two holes either square or rectangular at lower side for providing wooden flukes and a circular hole at the apex for tying cable. The Indo-Arabian anchors have been diversely named as 'Arabic or Protograpnel' (Frost,

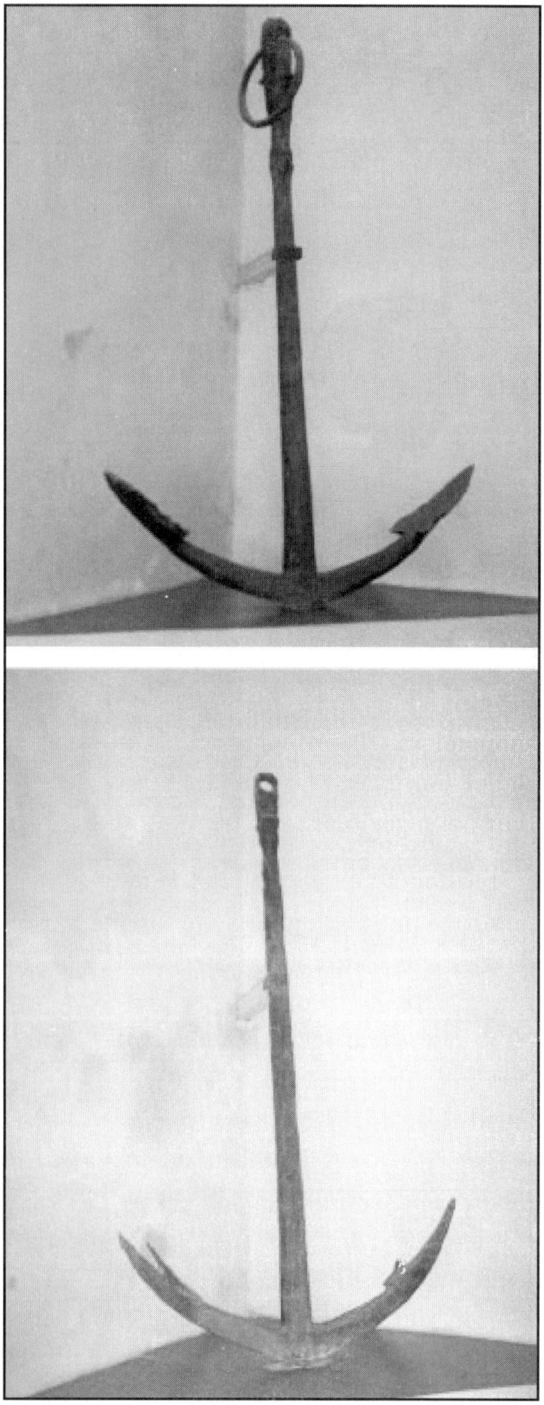

Fig. 68: Iron anchor displayed at the Archaeological Museum, Old Goa (Scale in 5 cm divisions)

1985), 'Arabo-Indian' (Souter, 1998), and 'Arabo-Indian grapnel' (Raban, 2000). However, in India these stone anchors have been termed as 'Indo-Arabian' because they were influenced by Arab mariners and similar in shape with those found in the Persian Gulf and Red Sea countries. Generally Indo-Arabian stone anchors are heavier and larger than the composite type of stone anchors. However, composite stone anchors are not reported from Goa waters but their distribution is very limited only to Gujarat coast of India. However, accidentally one composite stone anchor was found little away from the dockyard of Vijaydurg and some composite stone anchors also were found at Sindhudurg fort, along the Maharashtra

Fig. 69: Location of stone anchor find sites in India

coast which have been used on the platform of the jetty. These are the farthermost composite anchors so far reported from Indian coast. In addition to this these types of stone anchors have been recovered from Galle harbour of Sri Lanka.

Onshore and offshore explorations along the coast of India (Fig. 69) have brought to light numerous single hole, composite, Indo-Arabian and ring stone type of stone anchors from Aramda, Dwarka, Bet Dwarka, Miyani, Ghumli, Kodinar, Visawada, Somnath (Gaur et al., 2001; Gaur et al., 2002; Gaur and Sila Tripati, 2006; and Sundaresh et al., 1999), and from other sites of Gujarat namely Gogha, Hatab and Gopnath (Gaur and Bhatt, 2008), Sindhudurg, Padmagad, (Sila Tripati and Gaur, 1997), Vijayadurg (Fig. 70) (Sila Tripati et al., 1998a and Sila Tripati et al., 1998b), Kannur (Sila Tripati et al., 2005b), and Rameswaram, Kursadi Island, Mandapam, Periapattinam and Tuticorin of Tamil Nadu (Athiyaman, 2001), Dabhol (Gaur et al., 2009), Minicoy Island (Kapitan, 1994; and Sila

Fig. 70: Various types of stone anchors noticed at Sindhudurg, Vijaydurg and Padmagad

Tripati, 2009) and Chilika Lake (Sila Tripati and Patnaik, 2008) (Fig. 71 and Table - 4). Amongst the three Indo-Arabian stone anchors retrieved from Goa waters, two of them do

Name of the sites	Composite Anchors	Indo-Arabian Anchors	Ring stone Anchors	Single hole Anchors	Total
West Coast					
Bet Dwarka, Gujarat	13	7	1	------	21
Aramda, Gujarat	2	1	------	------	3
Dwarka, Gujarat	35	63	24	1	123
Miyani, Gujarat	2	6	4	------	12
Visawada, Gujarat	10	2	1	1	14
Kindar Kheda, Gujarat	1	------	------	------	1
Srinagar, Gujarat	1	------	------	------	1
Ghumli, Gujarat	------	------	1	------	1
Somnath, Gujarat	6	2	35	------	43
Kodinar, Gujarat	1	-----	-----	-----	1
Mithi Virdi, Gujarat	------	4	------	------	4
Gopnath, Gujarat	------	1	------	------	1
Ghogha, Gujarat	1	18	------	------	19
Hathab, Gujarat	------	4	------	------	4
Dabhol, Maharashtra	------	4	------	------	4
Vijaydurg, Maharashtra	1	23	------	------	24
Sindhudurg, Maharashtra	3	5	------	------	8
Baga, Goa	------	1	------	------	1
Sunchi Reef, Goa	------	1	1	------	2
Grande Island, Goa	------	2	------	------	2
Kannur, Kerala	------	1	------	------	1
Minicoy, Lakshadweep Is.	------	1	------	------	1
East Coast					
Chilika Lake, Orissa	4	------	------	1	5
Kottapatnam, AP	------	------	------	1	1
Manapad, Tamil Nadu	------	------	------	5	5
Kursadi Island, Tamil Nadu	------	1	------	------	1
Pumorichan, Island Tamil Nadu	------	1	------	------	1
Vedalai, Tamil Nadu	------	1	------	------	1
Periapattinam, Tamil Nadu	------	------	------	1	1
Tuticorin, Tamil Nadu	------	1	------	------	1

Table: 4 Various types of stone anchors have been discovered along the Indian coast (North to South).

not possess upper hole hence their use is enigmatic. Moreover, similar kind of stone anchors without having upper hole have been reported from Dwarka, Bet Dwarka, Tamil Nadu coast of India, Kilwa Kisiwani and Mogadishu on the east African coast (Chittick, 1980) as well from Oman. In all probability, these anchors might have been used as mooring bits, either unfinished or reworked after breaking at the apex. One of the Indo-Arabian stone anchors of Grande Island resembles in terms of their shape, size and chiselling pattern those found in Kannur, Kerala and Mogadishu, East Africa (Fig. 72). None of these anchors have upper hole. Either, these anchors were manufactured at one place, might have belonged to one mariner or made of same type of

Fig. 71: Stone anchors discovered at Tamil Nadu, Dabhol, Minicoy Island and Chilika Lake

Fig. 72: Stone anchors from Mogadishu Museum (East Africa), Grande Island (Goa) and Kannur, (Kerala) show uniform chiselling marks

rock which was suitable for such type of chiselling. In order to understand the introduction and distribution of Indo-Arabian stone anchors in Goa as well as other coast of India, it is crucial to know the origin of this type of anchor. So far no evidence is available as to how and where the Indo-Arabian type of anchor originated but the excavations at Siraf in Iran (Whitehouse, 1970) on the Persian Gulf have brought to light two fragmentary stone anchors (Fig. 73) from the stratified layers datable between 8th and 11th centuries AD. The recent past explorations have yielded Indo-Arabian stone anchors from Dibba, Dhofar,

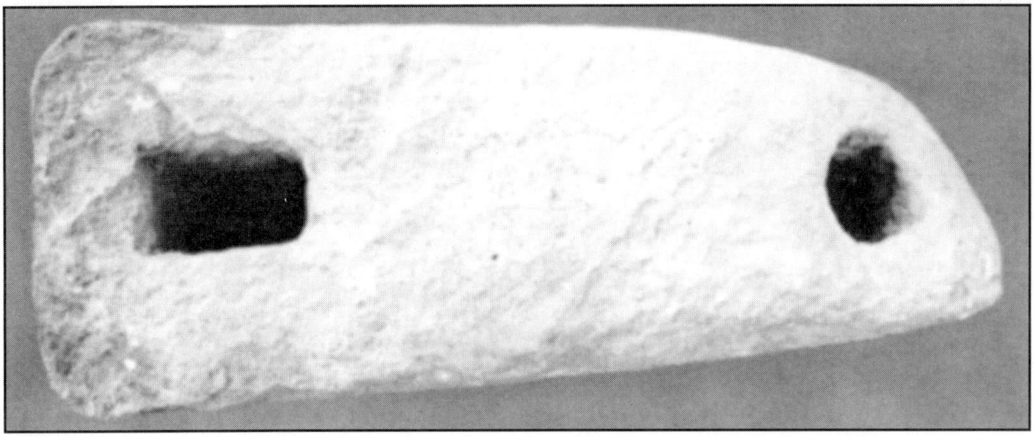

Fig. 73: Indo-Arabian type of stone anchor from Siraf

Masirah, Oman and Qalhat (Vosmer, 1999; and Agius, 2005). If the earliest date of Indo-Arabian stone anchor is accepted between 8th and 11th centuries AD, during these periods Arab mariners were getting more powerful and maritime trade was considerably growing in the Indian Ocean region. The country of the Arabs is surrounded by sea on three sides; consequently they had been enthusiastic tradesmen. Arabs had utilised Bahrain, Amman, Hazarmut, Yemen and Hijaz port which lie along the Indian Ocean, Red Sea and Persian Gulf for their sea borne trade. Among other items Arab horses were in great demand in India and Arabs were efficient horse dealers. Arab traders visited coastal ports of western India and took away the Indian goods such as sandalwood, camphor, cloves, nutmeg, coconuts, ivory, pepper, lead, etc to Iraq and Arabia. With the expansion of the Arabs under Abbasid dynasty, the Red Sea and Persian Gulf came under their control. During the time of Ibn Battuta (1304-1368 or 1369 AD) Indian ships coming from Thana, Quilon, Calicut, Kollam and Mangalore called at the port of Aden. As the Arabs were dominating in the sea trade, the Indian ships could hardly venture to go further west to the Red Sea. The western trade was controlled by the Arab and Persian traders hence Indian traders kept themselves confined to coastal trade. However, later on Marco Polo (1254-1324 AD) has mentioned that Qalhat was an excellent port. Every year several ships sailed from Hormuz and Qalhat to Indian coast with horses, dates, pearls and salt. They returned from India along with cloth, metalwork, spices and rice and these items were then traded in Persia and other parts of Arabia. Even, Indian vessels arrived with goods and in return they carry fine horses. Marco Polo states that the Pandyan port Cail (Kayal) was associated with horse trade from Arabia (Hourani, 1995; and Chakravarti, 1991).

During the 16th-17th centuries, horses were mostly brought to India from Arabia, Persia and Ormuz via sea routes and different types and sizes of ships such as *Baqueiras,*

Cotias, Tanos, Terradas and *Zambucos* were engaged in transportation of horses (Chauhan, 1984). Probably these ships carried Indo-Arabian type stone anchors. This is further

Fig. 74: Arab merchants shown on a dhow with horses on board in a 17th century mural painting in southern India

confirmed by a 17th century mural on the interior of Sri Narumpunadasvami temple wall of Tiruppudaimarudur, Tamil Nadu (Fig. 74) in which horses are shown on board along with Arab mariners (Deloche, 2009; and Guy, 2004). Moreover, these stone anchors have been reported wherever Arab boats called along the western Indian Ocean and the Red Sea coast. These anchors continued till the introduction of iron anchors by the Portuguese in Indian waters. The finding of several stone anchors across the Indian Ocean region indicates that there was an increase of volume of maritime trade during the medieval period. Moreover, it was revealing to learn that local seamen and fishermen of neighbourhood of Qalhat still used stone anchors in middle sized dhows in relatively modern times (Agius, 2005). The stone anchors of the Lone Mushroom wreck datable to 13th-14th centuries AD along the Red Sea coast having grooves at the base, probably meant for retrieving the anchor in case trapped in between rock. It could be the makers mark (Raban, 1990). Stone anchors having such type of grooves have not been reported either in the Indian or along the Western Indian Ocean coast. The Indo-Arabian stone anchors with the remains of wooden fluke have been reported from Galle harbour, Sri Lanka. The radiocarbon date shows that the wood is 430 ± 80 years BP (Souter, 1998). Moreover, in recent past Indo-Arabian stone anchors with wood remains have been found at Dabhol in Maharashtra coast. The wood samples were analysed using radiocarbon methods to determine the date of the wood. The calibrated date of the first sample ranged between 654 and 524 years BP, whereas the date of the second sample is placed between 653 and 537 years BP. The stone anchors of Goa waters are not found from any harbour sites or with a datable context. It appears that mariners might have dropped these anchors probably during rough weather, storm or in any other emergencies which were trapped in the rock and mariners could not retrieve those anchors therefore they left them. So far none of the Indo-Arabian anchors have been found prior to the 8th-11th centuries AD and the stone anchors of Sunchi Reef and Grande Island could be dated to the medieval or later medieval period. Though these stone anchors were used in India till 20th century, recorded documents namely the Portuguese *Cartaz* of 1704 AD mentions the use of stone anchors along with iron anchors in Indian waters (Shastry, 1973). The ring stone anchor of Goa resemble other ring stone anchors found in Dwarka, Bet Dwarka, Somnath, Minicoy and other sites of India as well as from the Persian Gulf, Gibraltar, Maldives and Mediterranean Sea but they differ in period. In India ring stone anchors were reported fewer than Indo-Arabian anchors. Once mariners experienced with a type of anchor which was found suitable for anchoring the boat, they used that variety and it continued for centuries. The finding of various types of stone anchors in Indian waters shows that the mariners adopted all varieties of anchors for the safety of their vessels. Arroyo (1975) mentions that several ring stones with circular hole have been reported from the Mediterranean Sea (Fig. 75) and these were ancient type of stone anchors which

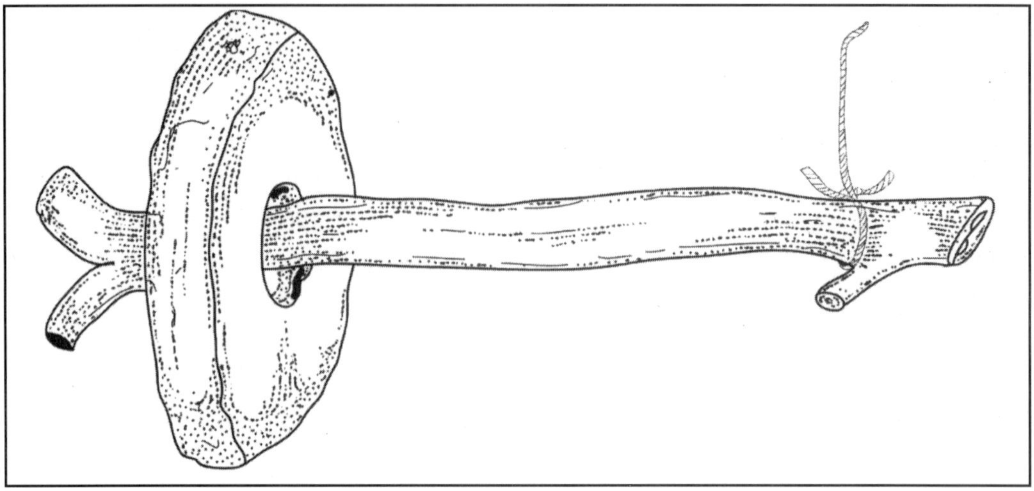

Fig. 75: Ring stone anchor found in the Mediterranean Sea

originated in the Mediterranean Sea and in later period the Arab mariners used these anchors in their waters. In the absence of tangible evidence it is very difficult to firmly propose that a particular type of anchor originated in a particular place. In India ring stone anchors were recovered from Dwarka waters in 1986 and onwards, then in Somnath waters. Even similar types of ring stone anchors have been reported from Oman (Vosmer, 1999). As Arroyo has suggested ring stone anchors originated in the Mediterranean Sea but during that time these anchors were not reported from the Indian Ocean region, hence now it becomes more essential to have a comparative study of these anchors.

The ring stone anchors were used since the 7th century BC as is evident from the logo of the Nautical Archaeology Society (NAS), UK. The logo has been copied from a bichrome painting from a Cypriot jug which measures 16 cm high datable to the 7th century BC. A ship has been portrayed on the jug showing two oars propelled from the stern by a helmsman. Another crewman is lowering a ring stone type of stone anchor (Fig. 76) from the bow. On the other hand it can also be interpreted that the crewman in the bow must be casting the ring stone anchor and he is in the act of climbing aboard with the anchor (Fenwick, 2000). The ring stone anchor depicted on

Fig. 76: Bichrome painting depicts a ring stone anchor on the 7th century BC Cypriot jug.

the logo is tied with a rope, without having the wooden stock and the size of the hole appears to be much wider. In this perspective it is difficult to understand whether wooden stocks were used or not. From the painting it appears that ring stone anchors were used from 7th century BC onwards, if not earlier. The lack of evidence in Indian context deprives us from understanding the earliest date of use of these types of stone anchors; however these anchors continued till the 20th century. Despite the introduction of iron anchors, stone anchors were used by the local mariners for centuries in India. Probably, the local fishermen or sailors who were navigating in local waters used the left over stone anchors for some duration and those were used as lintels of door, parapet wall, door sill, paving and mooring bits (Sila Tripati et al., 1998a). Similarly a broken stone anchor has also been used as architectural feature in the city of Qalhat datable to the 10th-16th centuries AD (Vosmer, 1999).

Evidences show that under the European influence iron anchors came into use in the 17th century, however, prior to that no information of use of iron anchors in Indian context. The text *Farhang-I Jahangiri* composed during the reign of Jahangir (1605-1627 AD) refers to '*langar*' as made of iron used to secure vessels and prevent from moving away. Another lexicon, *Bahar-I Ajam* compiled during the 18th century also mentions '*langar*' made of iron which was attached to the vessels. It appears that the production of iron anchors started in India in the second half of the 17th century which has been referred by Bowrey (1905). Further, the earliest evidence of the use of iron anchors in Indian waters comes from two boat paintings drawn by Bowrey in the 17th century 'Purgoo' and 'Patella' in which European type of two armed anchors have been depicted (Qaisar, 1982). From the shape and size of iron anchors it appears that the iron anchors of Sunchi Reef, the anchors of Archaeological Museum, Old Goa, anchors with wooden stocks off Aguada bay and other anchors of Aguada bar belong to the Admiralty Long Shanked anchor. Another almost identical iron anchor has been retrieved from the *Santo António de Tanná*, 1697 shipwreck off Mombasa, Kenya. This anchor was manufactured in Bombay probably in 1680.

Fig. 77: Iron anchor recovered from the dockyard of Vijaydurg

The wooden stock is missing and traces of the flukes and anchor ring remain almost destroyed by corrosion. Presently, the anchor is housed at Fort Jesus Museum, Mombasa (Curryer, 1999). Another iron anchor having similar shape and size with a 3.50 m long shank has been found at Vijaydurg dockyard and is presently kept in the Lal Bahadur Shastri Nautical and Engineering College, Mumbai (Fig. 77). In the beginning of the 18th century, Europeans started using the long shank iron anchors. The iron anchors of the Dutch, Danes and Swedes were five percent longer than the French iron anchors. However, the English started using 'Old Plan Long Shanked' anchors which were five percent shorter than that of the French and were straight-armed. Though there were some minor differences but they looked alike. These anchors had wooden stocks, which were nailed and in some cases metal brackets were provided for tightening the wooden stock (Rubin, 1971). Wooden stocks were added to iron anchors to provide flotation and buoyancy to anchors and to prevent damage to the anchors when hitting the rocky bottom. In some of the anchors the length of the wooden stock and the shank is equal. While the wooden stocks remained unchanged in use, in the wooden hulled vessels and the iron stocks were adopted in iron and steel hulled vessels (Upham, 1983). The wooden stocks of the iron anchors help one of the flukes to sink into the seabed, which helps to bring the ship under control; these wooden stocks are transverse beams. In these types of anchors one of the flukes sinks into the seabed under any circumstance and stock remains on the seabed in horizontal position (Kimpton, 1992). The smaller size iron anchors with flexible iron stocks which have been recovered from Aguada bar may belong to an Admiralty pattern anchor. This type of anchor with its improved construction and superior quality iron represented the final development of the fixed arm anchor. The function of flexible iron stock is different from wooden stocks. However, iron stock anchors came to limelight in 1807 AD. These types of anchors were introduced after 1841 by Admiral Sir William Parker and continued for longer period. In the 18th century iron stocks used for small anchors later on was used for larger anchors. The centre of curvature of arms is 1/3 distance up the shank from the crown. Its proportions of three to one were the same as those of the old long shank anchor. Their shape is elliptical and made of wrought iron with curved arms (Upham, 1983). It was noticed that iron anchors were frequently broken because of weight, poor quality of welding, straight arms, manufacturing defects and inferior quality of iron used for manufacturing the anchors. During the subsequent years, Richard Pering's improved the shape of iron anchor and recommended use of good quality of iron. The arms were then curved slightly instead of being straight. The form of the shank was slightly flattened at the sides instead of being round and the arms were oval. The length of the shanks and arms of the anchors were changed and these anchors were known as 'Pering's Improved anchors' (Cotsell, 1856).

Generally ships carry five to eight anchors depending on the size of the ship and the sheet or the best bower anchor is used only in emergency (Evan and Nutely, 1992). Sheet anchor was carried on the aft of starboard and used when bower anchors failed or dragged.

Conclusion

Stone anchors have been recovered from different parts of the world. In the absence of tangible evidence it is difficult to attribute a particular place for the origin of specific type of anchor. However, it has been generally believed that 'Indo-Arabian' types of stone anchors were introduced in Indian waters by the Arab mariners and they found this type of anchors suitable for coral and rocky seabed. The seabed of the Indian Ocean region falls under that category and mariners used these anchors for centuries together. Though Arab mariners had trade contacts with the Mediterranean Sea, Indo-Arabian stone anchors are not accounted beyond the Indian Ocean, Persian Gulf and Red Sea. Though Indo-Arabian stone anchors were associated with Arab mariners, the anchors reported from Indian waters had any direct relation with Arab traders or not is yet to be ascertained. In the subsequent years, Indian mariners found them suitable in the Indian Ocean region, hence they copied them and used for centuries. Further, these anchors provide evidence of maritime contacts between the Arabs, Red Sea and East African coast. Ring stone anchors have been recovered from the Indian Ocean, Persian Gulf, Gibraltar and Mediterranean Sea but they differ in period. In India more Indo-Arabian anchors were reported than ring stone and composite anchors. Once mariners found a type of anchor suitable for anchoring they adopted it and used it for longer period. The finding of various types of stone anchors in Indian water shows that the mariners adopted all varieties of anchors for the safety of their vessels. The finding of stone and iron anchors in Goa waters indicate a change from old technology and adaptation of new technology in the transition from stone to iron anchors. The admiralty type of iron anchor was used between the beginnings of the 15th century and remained typologically unchanged until the 19th century. Wooden stock is lighter than iron stock, hence earlier to iron stock anchors wooden stocks were used. Iron stocks were introduced to iron anchors at the end of the 18th century AD.

8

Explorations off Aguada Bay; Baga; Grande Island; Old Goa; Gopakapatana

Aguada was a harbour for providing safe anchorages to ships but every year Portuguese virtually used to close the harbour for navigation during the Southwest monsoon owing to high winds, rough sea and formation of sandbar in the estuary of the River Mondovi. The harbour had a lighthouse and many structures related to shipping and navigation. Within its preview there were some fountains and wells which were fuel and water supply point for the vessels inward and outward voyages to Portugal. During the Portuguese rule, ships were anchored off the Aguada Fort and only during high tide they sail to Old Goa crossing the Aguada Bar. 'Agoada' name has been derived from 'água' and means water or 'watering place' (Fonseca, 1986). Archival records mention that a number of Portuguese ships have been wrecked in the Aguada Bar. The most probable cause of wrecks on the bar appears to be due to inadequate knowledge about the seabed topography and unskilled navigation. The Aguada bar was referred to on Portuguese hydrographic charts as *Barra de Agoada.* The water depth at the bar is less than three metres and narrowed down the river mouth. The coastal processes of this region have created a narrow entrance in the river in which ships navigate and that is possible because of river flow and tidal circulation. The bathymetric survey profiles show that there is no topographic change during more than 100 years. Topographically this region has been considered as high energy zone. Dredging was carried out to deepen the navigational channel. In the course of dredging, iron anchors have been collected which are displayed at Goa State Museum. The shallow uncharted depth on the bar channel could be the cause of most of the shipwrecks in this area. Local fishermen many times inform that fishing nets generally entangle at '*Satabahu*' (where the water depth is fourteen metres) off Aguada. A two-fluked admiralty type of iron anchor with a nailed wooden stock was recovered in this area (Fig. 64). This anchor is the first of its kind reported so far in Indian waters. Similar types of anchor were used between the beginning of the fifteenth and the eighteenth century. While iron stocks were introduced at the end of the eighteenth century, wooden stocks continued in use on wooden hulled vessels. Preliminary survey was carried out behind Aguada Fort to locate the remains of shipwrecks. The dive area is full of rocks, at one location a square block was found but due to poor visibility further details of the block could not be obtained. The seabed of the dive area was hard and visibility was very poor. Similarly, exploration was also carried out opposite to circular bastions of Aguada fort to locate either shipwrecks or iron anchors. The whole survey area is full of rocks and fine silt. As exploration was confined to a very limited area no archaeological artefacts were found during survey. GPS positions of the site were also obtained (Sila Tripati, 2002a). More recently, an Admiralty long shanked iron anchor was found opposite to the lighthouse off Aguada bay in 11 m water depth in January 2008. Subsequently, it was retrieved. The finding site of the anchor is full of rocks and fine silt. The anchor is intact and its surface is uneven. A thick layer of coat has been applied on its surface. The coating layer on the wooden stock has been washed out, whereas; the coat can be seen on the flukes, shank and nails of the wooden stock. Even, underwater explorations carried out behind Cabo to locate archaeological remain if any. The whole area is full of laterite and almost every laterite boulder has crevices. In some places their height is 5 to 6 m. The seabed is full of coarse sand and dead shells. The exploration did not yield any archaeological findings (Sila Tripati, 2006b).

Underwater explorations off Baga

Baga is located approximately twelve to fifteen km to the north of Marmagao Bay. The bathymetric profiles show that the topography of the area is stable and smooth. Topographically the region is a low energy zone covered with soft mud. In 1989, explorations were carried out off Baga to locate shipwrecks in 3 to 6 m of water depth. The sea floor in the near shore region is rocky and thickly covered with sponges, while in 5 to 6 m water depth coarse sand was encountered. Fine sand and silt occur in 7 to 8 m off Baga; beyond this region the area is full of mud and clay. The metal detector survey off Baga confirmed the existence of a metallic wreck in 6 m water depth opposite the seminary. Visual surveys confirmed this inference. The wreck is buried in sand and sediment and only a little portion is visible. Due to turbid water, it was difficult to know the extension of the wreck site and the remains could not be recorded due to murky water and poor visibility. The ship might have struck against the submerged rocks and sunk. Further, exploration brought to light a two holed stone artefact with a rectangular cutting on the top which appears to be a pyramidal stone anchor. The stone anchor was retrieved for study. But the stone anchor looks like a broken Indo-Arabian type anchor at the first sight. In general pyramidal stone anchor used to have a top hole which goes up to the rope hole, but this stone anchor does not have such perforation. The provenance of the stone anchor is difficult to be known because no datable artefacts have been found along with it. Like other types of stone anchors, pyramidal stone anchors have not been found in profusion in the Indian Ocean. These types of stone anchor have been reported only from limited sites; this is the first of its kind found in India so far (Sila Tripati, 2002b).

Underwater explorations off Grande Island

The explorations off Grande Island yielded a steel hulled cargo vessel with cabin, hatch and bridge (Fig. 78). The cabin is ransacked, its lower deck and hull, lie covered by thick layer of sand. The vessel lies tilted towards the star board side and small portion of the deck

Fig. 78: Shipwreck remains off Grande Island

covered by coral is visible. The cabin is visible during low tide (Thakkar, 1991a). The mast and ladder are lying on the seafloor. These are scattered over a large area, probably the wreck is less than a century old, but no records are available about the shipwreck. It appears that the ship might have sunk in a storm after it lost anchor and drifted towards the rock near the shore of the island. It is informed by the local divers that many parts of the ship are lying on the southern side of the island. As the ship is ransacked the length, width, flag, destination and other related details cannot be ascertained.

Underwater explorations off Old Goa

References reveal that the hillock of Old Goa on which 'Our Lady of the Rosary' church is constructed has been known as holy hill. From the NW corner of the hill, one can clearly notice Raibandar and the River Mandovi. Probably the River Mandovi was flowing close to the hill during earlier times. It appears that Afonso de Albuquerque ordered to fire guns against the forces of Adil Shahi from this hill. During the Portuguese period, Old Goa on the bank of River Mandovi was the capital and port city and ships were constructed and repaired at the dockyard of Old Goa. With this background it was decided to undertake underwater explorations between Captain of port jetty and the boat yard of the Pentagon Engineers to locate the dockyard remnants, Portuguese shipwrecks if any and remains of maritime activity of Goa. The riverbed is smooth, covered by fine silt, there is no outcrop of rock but laterite boulders are present and the visibility is fair. Some places compact clay mixed with coarse sand with live shells is also noticed (Sila Tripati et al., 2006b). Diving and metal detector surveys were carried out and some sherds of Chinese ceramics, celadon and indigenous pottery was found during airlifting. Extensive airlifting was carried out in 6 locations at more than 1 m depth in the riverbed of Mandovi. As the Celadon and indigenous potsherds are small and highly rolled, and the main shapes are bowls, dishes, jars and pots in the collection and rendering, it is difficult for identification of designs and paintings. The pottery is mostly well fired and fabric is medium to thin. Painting is not visible on indigenous pottery, probably washed out due to lengthy immersion in water. The Chinese ceramics belong to the hard paste variety and are painted in blue on white. Analogous Chinese pottery is found in the Old Goa region which suggests that pottery was coming to Goa and Maharashtra coast through trade by the Mughal, Portuguese, English and Dutch. Metal detector survey revealed some clamps and shovels which were buried in sediment of the riverbed. In the river, current action is very strong (Sila Tripati, 1999).

Explorations at Gopakapatana

Onshore explorations and trial scraping were carried out to trace the earliest evidence of Gopakapatana. In order to ascertain the continuation of Rajvidhi (Fig. 79) towards the shore, trial pits were laid at the junction of the sea and road of the village Donda. The pits

Fig. 79: 'Rajbidi' or 'Rajvidhi' of the Kadamba period at Gopakapatana

brought to light laterite blocks and laterite blocks with lime mortar and gravels with reddish sand. These findings could be the part of the Rajvidhi of the Kadamba and Shilahara period (Gaur, 1992). In the course of exploration, laterite stone walls running perpendicular and parallel to the River Zuari are noticed. These walls run between Gopakapatana and Agasaim. Well dressed laterite blocks of different sizes have been used in the construction. In some places these walls are buried in the sediment and also disturbed. The longest wall of 1.5 km in length has been noticed at Gopakapatana which gets exposed during low tide. Trial diggings were carried out very close to stonewall to understand the nature and purpose of these structures. Several bastions at regular intervals along one of the walls and six stone steps leading towards the shore side have been observed (Fig. 80). The steps are attached to the main wall, but due to seepage of water further excavations could not be carried out to reach to the lowest step. All these stone structures lie very close to the broad royal road of the Kadambas that extended from Gopakapatana to Ela (Old Goa). The explorations at Gopakapatana so far has not yielded any datable findings that could have supported to determine how old these structures were (Sila Tripati et al., 2006b; and Sila Tripati et al., 2008). It is very difficult to state whether these structures belonged to Gopakapatana port or not. Similar structures have been noticed along the coastal region of Goa. Whether these structures of Gopakapatana belonged to Kadamba period or not still remain unanswered.

Fig. 80: Stone steps in the wall leading towards the shore at Gopakapatana

Large scale explorations and excavations would help to understand these stone structures and the reasons why they were constructed. However, the artefacts, which have been displayed at Pilar Museum, have been collected from Gopakapatana and adjoining region, show that Gopakapatana may have been an ancient port of Goa (Gudigar, 2000).

9

Studies on the Elephant Tusks and Hippopotamus Teeth collected from Sunchi Reef Shipwreck

Prior to the expedition of Vasco da Gama (b. 1460 and d. 24 Dec 1524) to India, a series of Portuguese navigators had vainly attempted to sail to the East. Their voyages terminated in and around the African coast. On 08th July 1497, with instructions from Lisbon, Vasco da Gama left for the East with the flagship *São Gabriel* and the *São Raphael* and the caravel *Bérrio.* After making landfalls at St. Helena Bay, Mossel Bay, Mozambique, Mombasa and Malindi he eventually reached Calicut with the assistance from an Indian pilot Ahmad Ibn Majid, who guided him across the Arabian Sea to India. Initially there were resistances against the Portuguese both from the locals as well as rulers. Subsequently, agreements were signed between the Portuguese and local rulers of India for trade and commerce and a number of trading stations were established along the west coast of India. The Portuguese monopoly over the maritime trade continued for more than a century and their ships plied the high seas along the shores of the Indian Ocean region without competition until the arrival of the British, the Danes, the Dutch and the French who also established their trade centres in India. The basic pattern of trade of the European countries was to bring soldiers, trade goods, administrators, furniture, war material, building material etc. to the region and on return to carry spices, porcelain, cloth, silk, rice, timber and finished manufactured products. In course of time, vessels from many European countries have been wrecked in Indian waters with a greater proportion of Portuguese ships wrecked in Goa waters. As these ships carried valuable cargo and artefacts, attempts were often made to salvage shipwrecks, sometimes successfully. Over the years Portuguese shipwrecks have been explored and excavated off Africa, India and Portugal. Most of the cargo has left no tangible traces of the spices, tea, cloth, silk, rice and other organic substances that they would have carried; those are most likely to have been completely degraded or to have been washed away. Exceptionally pepper remains have been recovered from the *Nossa Senhora dos Mãrtires* 1606 shipwreck off Lisbon. Other cargo such as guns (cannons), anchors, ceramic and stoneware sherds as well as building blocks still lie on the seabed as these are of little commercial value to the salvagers and treasure hunters. The record of Portuguese shipwrecks started in Indian waters after the arrival of Vasco da Gama at Calicut on 17 May 1498. Archival documents reveal that several Portuguese ships wrecked in Indian waters between the 15th and 18th centuries AD, the majority of them was due to unskilled navigation, warfare, hidden rocky reefs, woodborers and natural calamities.

Sunchi Reef lies between Cabo headlands and Marmagao (Fig. 9). The whole region consists of laterite shoals which extend in north-south direction, and separate Marmagao Bay from the Arabian Sea. Due to shallow water the strong current prevails, and severely hamper the survey work in the region. The seabed is covered with shelly sand and lithified sediment (Wagle et al., 1988). Extensive and thorough search in and around Sunchi Reef led to the recovery of sherds of storage jars, the barrel of a handgun, Chinese ceramics, elephant tusks, hippopotamus teeth, guns (cannons), an iron anchor, handgun, gun shot, bases of glass bottles, pieces of lead pipe and sheets, copper vessel and strap, stone and clay bricks and a number of dressed granite blocks (Sila Tripati et al., 2004; and Sila Tripati, 2004; and Sila Tripati et al., 2006a). These artefacts were either found in the laterite crevices or covered with a thin layer of silt or buried in the sediment. The guns and iron anchor however were lying on the laterite with prolific growths of marine organisms on their surface (Sila Tripati et al., 2001).

Ivory Trade between Goa and Portugal

Commodities like ivory, gold, ebony and some precious metals were brought to Goa in Portuguese ships in exchange for Indian spices, cloth and food grains (Boxer, 1969). The main items of export by the Portuguese were pepper, ginger and other spices from Malabar and Karnataka, cotton and silk from Bijapur, Bengal, Cambay and Golconda, cinnamon from Ceylon and mace, nutmegs and cloves from the Maluccas and Banda Islands. Similarly, ivory, gold, slaves, precious metals, stones, corals and other articles of trade were brought to Goa from Mozambique, Mombasa, Sofala, Angola and other places in Africa (Fig. 81). Ivory was the main trade item along with slaves and gold. The Portuguese generated huge revenue and wealth by trading with these centres, with ivory one of their most profitable

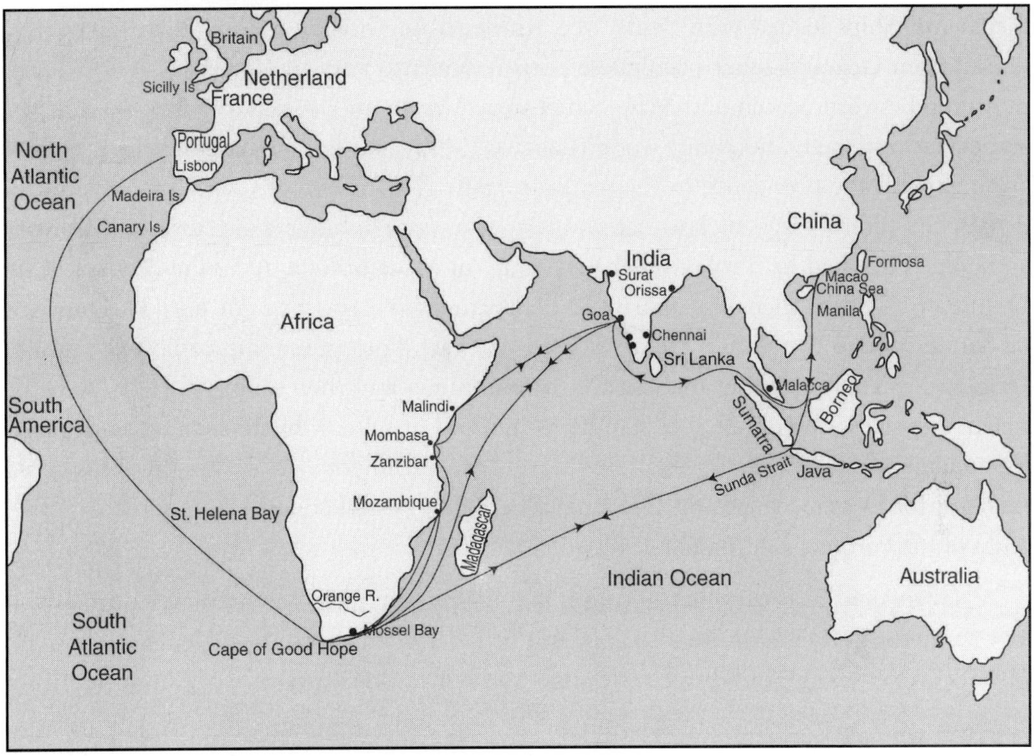

Fig. 81: Portuguese trade centres along the African coast, India and Macao, and routes followed by Portuguese ships

commodities. After receiving ivory from African countries those were sent to Cambay, Diu, Surat and other places in Gujarat along with arecanut, coconuts and drugs because artisans concentrated in this region were more skilled than those in other regions (Sila Tripati and Godfrey, 2007). They enjoyed a *de facto* monopoly on producing ivory goods and no one could compete with them. Even lead, pepper, cloves, ginger, nutmeg, ivory and other items were sent from Goa to Bahrain, Basra, Muscat and Sind (Bragança Pereira, 1937). The ivory was stored at the house of Manuel Moraes of Goa and subsequently sent to Cambay or Surat as per the particular contract. Even ivory imported by private merchants was carried under the Portuguese banner. The *nau São Miguel*, for example, sailed from Mozambique to Goa carrying 600 quintals of copper and eighty pieces of ivory (Stephen, 1997). Sometimes *naus* carrying ivory sailed directly from Mozambique either to Cambay, Diu or Surat since ivory was in great demand. Further, Portuguese records mention that a *nau* called *Nazaré* belonging to the king of Portugal sank at the bar of Goa in 1523 AD and some of the cargo such as ivory and copper were recovered by diving (Stephen, 1997). Textiles purchased from Cambay and Surat were sent to Malindi through the Hindu merchants. Ships loaded with beads, rice, textiles from Gujarat and materials for warfare set sail from Goa and other Portuguese ports in India to African countries every January, returning between March and May loaded with African goods such as gold, ivory, slaves, amber, tortoise shells and ebony wood (Lobato, 1997). During 17th century the *Estado da India* enjoyed the prevalence in the maritime trade. The *Vanias* of Goa, Daman and Diu despatched Indian cargo such as rice and textiles to Mozambique and returned with ivory and gold. They had firm role over the economy of Mozambique. After the death of King D. Jose in 1777 AD there was disarray in the Portuguese maritime trade until D. Francisco de Souza came to the rescue. He considered that each Portuguese ship could fetch 40000 *cruzados* per year by selling the cargo at Mozambique and then sailing for Goa with the remaining goods and other merchandise purchased at Mozambique such as gold, ivory, hippopotamus teeth, tortoise shells and slaves. After selling goods at Goa the ships could then sail for Daman, Surat and Diu from where they would collect all consumable goods and sail for Portugal (Shirodkar, 1988).

Portuguese records which indicate the volume of ivory imported to Goa between 1613 and 1661 AD are summarised in Table 5. It is evident from the table that more than 50% of ivory was imported from Mozambique but the quantity of ivory imported from

Year	Volume in kg	Amount invested in Xerafins	Imported from
1613	26691	79332	Mozambique
1614	23282	69246	Mozambique
1619	11641	17695	Mozambique
1620	2308	3476	Mozambique
1621	2622	3985	Mozambique
1625	1232	1870	Mozambique
1626	1311	1993	Mozambique
1644	944	1432	Mozambique-Lisbon
1646	44834	72996	Mozambique-Diu
1650	1229	1846	Mombasa
1661	55884?	84944?	Mozambique
1661	34642	52655	Mozambique

(Source: Ahmed Afzal, Indo-Portuguese trade in the seventeenth century 1600-1663, Gian Publishing House, New Delhi, 1991, p. 151).

Table 5 Quantity of ivory imported to India from African countries

other countries is not clear (Ahmed, 1991). It is generally believed that the ivory of the African elephant is much better in quality than the Asiatic because of its superior hardness, pale blonde transparency and ability to be more finely polished. African elephant tusks can measure up to 2 m in length with diameters of 9 to 11 cm and weights up to 90 kg while tusks of Asiatic elephants are smaller and lighter. Ivory and hippopotamus teeth were used for making of sculptures, statues, bangles, various types of handles for various usages, seals, cones, pyramids, buttons, clips, cylinders and toy animals in a wide variety of shapes and sizes. Besides ivory statues, inlay ivory figures were incorporated in chairs, tables, wardrobes, chests, sofas and boxes for keeping gold and jewellery. Indian artists were experienced with non-Christian art prior to the arrival of the Portuguese. Influenced by the Portuguese they started making sculptures and statues of non-religious, secular as well as profane artefacts of the Christian art made of stone, wood and ivory in Goa and elsewhere. Hippopotamus are large aquatic mammals that live in the wetlands of Southern, Western and Central Africa. Hippopotamus were once found throughout Africa but now they are extinct in northern Africa. Over the centuries people have hunted hippopotamus for meat and teeth. The enormous mouth contains large canine tusks which may be over 1.5 m long. The hippopotamus is also known as 'river horse'. The teeth of Hippopotamus are the hardest of all, retain their whiteness for a long time and do not change colour when cut and polished. Hippopotamus have enormous jaws that hold long curving teeth which consist primarily of enamel.

Composition, Structure and Properties of Elephant Ivory

Ivory is dentine, a non-cellular substance present in all teeth including those of the walrus, hippopotamus and sperm whale. Each type of ivory has its own characteristics and distinctive processes by which dentine is formed. African ivory is of the highest quality, hardest and most brilliant. It also tends to retain its yellowish-white colour better as it ages. The best African ivory comes from Gaboon, Mozambique and Zanzibar and as stated earlier, was much in demand by Indian craftsmen for making sculptures, bangles, inlay work and other artefacts. A transverse section of a piece of un-degraded African ivory shows its typical pale blonde transparency and the clear, sharp, engine-turned decussating appearance characteristic of elephant ivory. These markings are clearly evident in the transverse section (Fig. 82) of a piece of degraded ivory recovered from the Sunchi Reef shipwreck. As described by Penniman (1952), 'the longitudinal section shows straight continuous lines, and usually long thin parallel clouds which are whiter and more opaque than the material between them'. Indian ivory, on the other hand, is whiter, softer and more opaque, with a tendency to discolour and takes a less fine polishes than African ivory. When exposed to the light it tends to

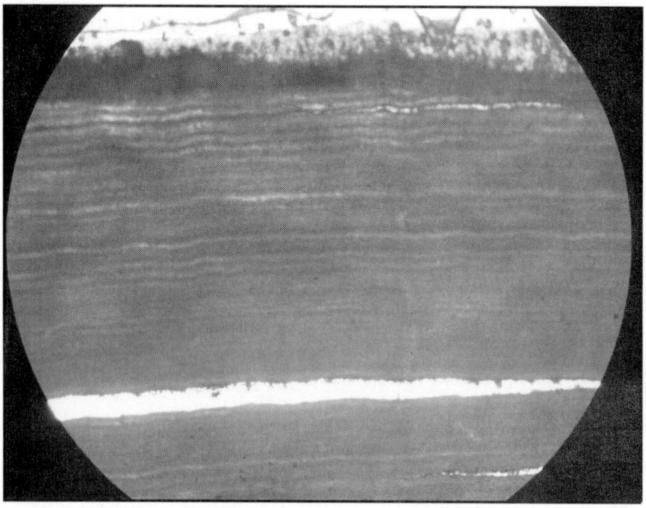

Fig. 82: Transverse section of an elephant tusk from the Sunchi Reef shipwreck site (25X).

become duller in appearance whereas African ivory often takes on a porcellanous sheen. The appearance of a cut section is creamy with a mellow sheen and the engine-turned decussating appearance on the transverse section is less distinct in the Indian ivory. As with African ivory the longitudinal section of Indian ivory also had the characteristic straight lines and long, thin parallel milky clouds. Fresh ivory contains two main components, a complex protein that permeates the structure and a mixture of minerals that provide strength. The ratio of inorganic to organic matter is about 65:35 in African elephant ivory (Webster, 1958). The principal organic component is Type I collagen, a protein with a triple helical structure that contains a high proportion of the amino acids, glycine, proline and hydroxyproline. The major mineral component is a non-stoichiometric, carbonate-containing apatite, similar to hydroxyapatite. Evidence from X-ray diffraction and infrared studies indicate that this mineral is best represented by the formula $Ca_{10} (PO_4)_6 (CO_3) H_2O$ (McConnell, 1965; and Brophy and Nash, 1968). Small amounts of calcium carbonate and calcium fluoride comprise the bulk of the remainder.

Chemistry and Deterioration of Ivory

The physical, chemical and biological characteristics of the burial environment and the period of interment largely determine the condition of recovered ivory artefacts, with microbiological activity critical in the early stages of deterioration (Nicholson, 1996). The breakdown of collagen leads to the formation of more soluble peptides and amino acids which are subsequently leached from bone (Garlick, 1969). This breakdown may be attributed to either microbiological attack or chemical hydrolysis. The latter mode of collagen deterioration is expected to be predominant in environments which inhibit demineralisation of the inorganic matrix (Child, 1995). Ivory's porosity enhances both exchange and deposition of minerals within the inorganic matrix. The breakdown of collagen increases the exposure of the hydroxyapatite crystallites and thereby promotes crystallographic change

(Sillen, 1989; and Pate et al., 1989). Hydroxyapatites are also known to undergo alterations such as anionic exchange with fluorine, cationic exchange with strontium, radium and lead and substitution of the phosphate moiety with carbonate (O'Connor, 1987).

Elephant tusks and Hippopotamus teeth and their state

The explorations off Sunchi Reef, Goa, brought to light eight elephant tusks of different sizes ranging from 65 to 32 cm in length and their circumference varies from 15 to 22 cm. These tusks were underlying in the coarse sand and covered with dead shells with only small portions observable above the seabed. All tusks are cut pieces except the longest one which measures 65 cm and has the distal end intact. The long period of immersion has produced degraded ivory that is soft, brittle and flaky. Similarly, nine distinct sizes of hippopotamus teeth, partly buried in the seabed to the east and northeast side of the guns were retrieved. On each tooth a straight cut-like depression is present, that is because of continuous rubbing (abrasion) between the lower and the upper jaw. The teeth of Hippopotamus are the hardest of all teeth that are used as ivory. They retain their whiteness for a long time and do not change colour when cut and polished. These teeth are composed of enamel, the outer 'glazed' surface and dentine, the hard inner substance. The white teeth have turned brownish in colour due to their prolonged exposure to marine environment. Presence of marine organisms on their surface also supports this view (Sila Tripati and Godfrey, 2007). After excavation the elephant tusks and hippopotamus teeth were kept in marine water for some days. Subsequently 5, 10, 15, 20 % of fresh water was added to them by removing the same quantity of marine water. Salt analyses were used to estimate the quantity of salt in the water, with desalination continuing until all detectable salt had been removed from the tusks and teeth. On completion of desalination the tusks and teeth were placed in fresh water, with the solution changed every two months or so. Each of the tusks has been bound with twines to retain flaking and splitting fragments. The ivory, which appeared to be highly degraded, was soft to the touch and discoloured. This discolouration was not uniform indicating that the ivory was differentially degraded. Previous work on bone and ivory had indicated that Fourier transform infrared (FTIR) spectroscopic analysis was capable of providing information about the inorganic and the organic components that make up these material types (Godfrey et al., 2002).

Experimental Techniques

FTIR spectra were collected using a microscope diamond cell accessory on a Biorad Excalibur FTS 3000MX FTIR spectrometer. Duplicate spectra were collected for each sample. Samples were taken from areas in which different colouration indicated that there may be differential degradation of the ivory. FTIR spectra were collected for samples that were identified as being 'black', 'inner', 'pink', 'rusty' and 'general'. Heights for the peaks centred at approximately 1650-1620 cm^{-1}, 1420 cm^{-1} and 1035 cm^{-1} were determined using the Galactic Industries Corporation GRAMS/386 software and a baseline of 1800-800 cm^{-1}. The amide I peak (1650-1620 cm^{-1}) and the phosphate peak (PO_4^{-3}), centred at approximately 1035 cm^{-1}, were used as the respective reference absorptions for the organic and mineral components of ivory while the peak centred at 1420 cm^{-1} was used as the carbonate (CO_3^{2-}) reference peak. The intensity ratios I_{1640}/I_{1035} and I_{1420}/I_{1035} were calculated for all duplicate samples to quantify the relative retention of organic material and the incorporation of carbonate, relative to phosphate, in the inorganic matrix respectively.

Results of Analysis

Data pertaining to the FTIR spectra are recorded in Table 6 as are FTIR spectra for modern, un-degraded African elephant ivory and for a typical piece of degraded Goa ivory (Fig. 83).

FTIR Spectral Data

Sample	I_{1640}/I_{1035}	I_{1420}/I_{1035}
Black	0.12	0.56
Inner	0.17	0.64
Pink	0.20	0.63
Rusty	0.42	0.70
General 1	0.16	0.66
General 2	0.14	0.56
Modern ivory *	0.62	0.16

* Godfrey *et al.*, 2002

Table 6. FTIR Spectrum data of Elephant tusks of Sunchi Reef and Mozambique.

All of the spectra of the archaeological samples differ from that of un-degraded ivory and provide information on the impact of immersion and diagenetic change. Characteristics of the FTIR spectra include the loss of substantial amounts of collagen, the retention of the hydroxyapatite inorganic framework and the incorporation of carbonate into the structure. The IR spectra of modern ivory (un-degraded) and Goa ivory (degraded) retrieved from Sunchi Reef, Goa were compared. The absence of peaks between 1540 and 1660 cm^{-1} in the latter is due to degradation of collagen while the stronger peaks at 870, 960, 1420 and 1450 cm^{-1} in the Goa sample may be attributed to its richer carbonate content due to incorporation from carbonate rich seawater. The IR spectrum of the hippopotamus teeth also resemble that of the Goa ivory as it too has very little collagen, but higher level of Carbonate. Evidence for the depletion of collagen from the archaeological samples is provided by absorptions in the 1700-1500 cm^{-1} region. For each sample the intensity of absorptions in the amide I region, centred at approximately 1640 cm^{-1}, is markedly diminished in relation to the characteristic phosphate absorption centred at approximately 1035 cm^{-1}. It is interesting that slightly more collagen appears to have been retained in the 'rusty' samples. Although no analyses were carried out to determine if in fact iron is present in these areas. The typical rust

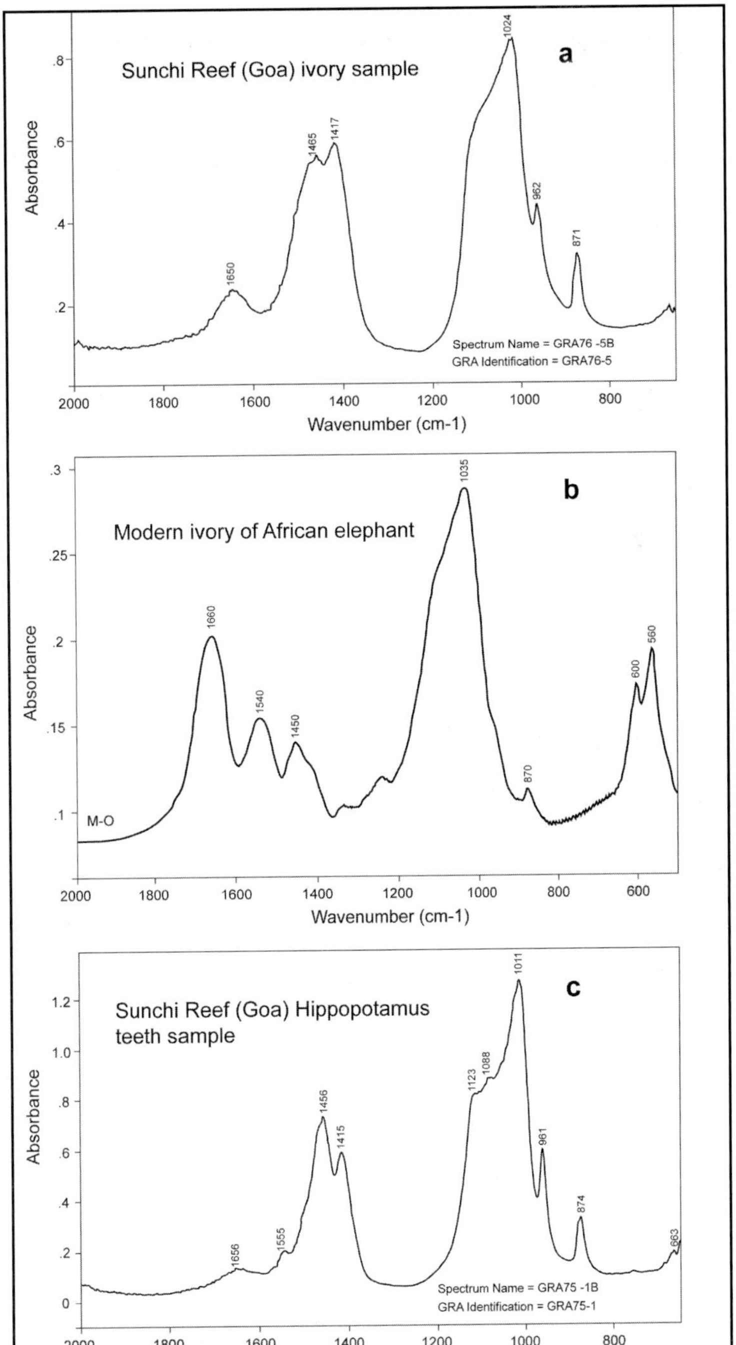

Fig. 83: FTIR spectrum of an elephant tusk and sample of modern ivory of African elephant and hippopotamus teeth from Sunchi Reef

colour indicates that this is likely. If this is the case, it would provide further evidence for the preservative effects of iron corrosion products on immersed collagen. In a recent study (Godfrey et al., 2002) greater amounts of collagen were found in the outer, iron-rich regions of a tusk fragment than in the iron-depleted, inner regions of the tusk. This tusk has been recovered after 350 years of immersion in a marine environment. Absorptions at approximately 1035 cm^{-1} and 960 cm^{-1}, typical of phosphate and 1460 cm^{-1}, 1420 cm^{-1} and 873 cm^{-1}, typical of carbonate in un-degraded biological apatites (Furedi and Walton, 1968) indicate that the carbonate-containing hydroxyapatite structure has been substantially retained in these materials. Interestingly, very high carbonate/phosphate ratios (I_{1420}/I_{1035}) for all samples indicate that carbonate substitution for phosphate has been a very significant part of the overall diagnosis of the ivory found off Goa. This finding, although contrary to the Western Australian tusk fragment described previously (Godfrey et al., 2002) is consistent with the proposal that an increase in the amount of secondary carbonate in the matrix may be related to the loss of organic matter (Pate et al., 1989). C14 dating of the ivory gave an age range 740 ± 130 yrs old with a calibrated age range of 740 to 560 yrs BP.

Discussion and Conclusion

Elephant tusks and hippopotamus teeth have been traded for thousands of years. The excavations of the Uluburun shipwreck off the southeast coast of Kas, Turkey yielded several artefacts including elephant tusks and more than a dozen hippopotamus teeth, dated to 1400 BC (Bass, 1996). The Portuguese maritime explorers turned towards India primarily in search of spices which were in great demand in Europe. Being emboldened by the discovery of the sea route to India the aim of the Portuguese was to obstruct as much as they could the trade between India and the Red Sea and between India and the Persian Gulf and to channel the trade of the East with Europe as well as Portugal. To achieve these goals they strove to build up Goa as a main entry point for all kinds of activities. The Portuguese mainly dealt with high demand goods which could fetch high profits with ivory, one of the most profitable of these. After discovering the new sea route to the East at the end of the 15th century, the Portuguese understood the importance of and high demand for ivory in India, importing it from African countries to meet this demand. The ivory tradition was not known to the Portuguese prior to this and it was a surprise for them to see the extent of developed ivory art in India. Although Indian ivory has been used for making miniature sculptures and artefacts since the beginning of the Bronze Age civilisation in India, it was unsuitable for making big sculptures because of its brittle texture, softness, opaqueness and tendency to discolour. Excavations at Harappa, Mohenjodaro and Chanhudaro for example have yielded ivory and other artefacts. The ivory comb shows the characteristic layered breakages and other artefacts have turned brownish coloured to pale dark (Penniman, 1952). Artefacts made of Indian ivory were smaller in size hence the Portuguese brought longer ivory from African countries for making bigger sculptures. With the passage of time the export of Indian textiles to the African countries increased concurrently with increased volumes of ivory heading back to India. Many ivory sculptures and statues are found in museums and private collections in India, Portugal, Mozambique, Brazil, Great Britain, France, Holland and their former colonies which had exported finished products from India through the port of Goa. The ivory from the wooded and humid regions of Africa was considered better for carving sculptures, statues, bangles and other objects of religious and secular art. Thus, despite the difficulties of using morphological features to unambiguously assign a provenance to ivory artefacts, these regions of Africa are the most likely source of the ivory used in such artefacts. It was not only the Portuguese ships that were involved in the ivory trade. In the mid 19th century the wooden hulled British ship *Child Harold* sailing from Bombay to London and carrying 1336 pieces of ivory, cotton, deer horns, pearl shells and cardamoms wrecked off Dassen Island, South Africa, in February 1850 (Turner, 1988). The Dutch East Indiaman *Vergulde Draeck* struck a reef close to the Southland about 120 km north of Perth, Australia in 1656 AD. The underwater explorations brought to light iron guns, cannon balls, ballast bricks, resin, slate, wooden and bone handles, comb, shoes, timer, pitch and 13 elephant tusks. It was presumed that these elephant tusks were exported from the Netherlands (Green, 1973). Similarly, the Danish-Norwegian ship *Fredensborg* involved in triangular trade (Europe, Africa and West Indies) wrecked off the Island of Arendal in Norway in 1768 AD carrying slaves, 927 kg of elephant tusks, hippopotamus teeth, cinnamon, tobacco and cotton. These elephant tusks and hippopotamus teeth were from Africa (Delgado, 1997). Two Hippopotamus teeth have also been recovered from the wreck of the *Santo António de Tanna* (Darroch, 1986). Unfortunately no elephant tusks were found on the *Santo António de Tanna.* Excavations of the VOC *fluit Risdam* (1727) near Mersing on the east coast of Malaysia have yielded tin ingots, lead ingots, 90 elephant tusks, Dutch bricks and storage jars. It appears that these tusks may have been of Southeast Asian origin (Green, 1986).

The engine-turned decussating appearance usually found on the transverse section of elephant tusks was not detected in the Sunchi Reef elephant tusks because these tusks are highly deteriorated and degraded. Likewise the usual features such as colour, hardness and opaqueness which can sometimes be used to differentiate between African and Indian ivory could not be used for these deteriorated samples because of the changes in these properties that occurred during prolonged immersion. As hippopotamus are now confined to only African countries, it is this region that is the original source of the hippopotamus teeth which were recovered from Sunchi Reef shipwreck site. The co-location of hippopotamus teeth of definite African origin with elephant tusk material indicates that this latter material is also likely to have been sourced from Africa. Further, as the largest quantities of ivory were imported from Mozambique, it is likely that this country was the source of the tusk material and hippopotamus teeth found on Sunchi Reef shipwreck site. The historical record and the co-location of hippopotamus teeth with the tusks therefore provide the most compelling evidence for Sunchi Reef tusks having an African origin. Although further investigations may shed new light on the origin of the teeth and the tusks, conclusive determination of their provenance will not be an easy task. DNA analysis, for instance, which could differentiate between Indian and African ivory is problematic, due to the deterioration of the organic compounds caused by prolonged immersion in the marine environment.

10

Use of Timber in Shipbuilding Industry: Identification and Analysis of Timber from Shipwrecks off Goa Coast

Indian texts and archival records mention that seaworthy vessels of various types were built in India for overseas trade, commerce and naval warfare. Before the use of steel in the shipbuilding industry, timber was used extensively for the construction of boats and ships. Panini (5th century BC) has mentioned that a large variety of timber was used in construction of different parts of a ship (Agrawal, 1953). India has a long maritime history but there is a very little information available on as to the number of ships built, their purpose, history and other details. Historical documents and archival records indicate that there were several wrecks off Goa, causes attributed to natural calamities like cyclone, collision against submerged rocks, reefs, sandbars, human errors and naval warfare (Mathew, 1988). In view of this, maritime archaeological explorations were initiated in Goa waters in 1989 with the objective to explore and study the remains of shipwrecks, causes of shipwrecks, their period and provenance. Over the years, explorations have been carried out in Sunchi Reef, St George's Reef, Baga and Aguada waters along the Goa coast (Thakkar, 1991b; and Sila Tripati et al., 2001; and Sila Tripati et al., 2002; and Sila Tripati et al., 2006b). An Admiralty Long Shanked type of iron anchor measuring (2.65 m) long shank with 2.53 m nailed wooden stock was recovered from the surrounding area of Aguada bay. The finding of timber from the shipwrecks provided information on the quality of timber used in construction of ships and anchors during that period. Analysis of timbers recovered from

St George's Reef and Aguada bay were carried out in order to understand the anatomical details of the timber, their state, species, dates and effect of woodborers on them.

Materials and Methods of analysis of timbers

The materials used for the study consists of small uncarbonised and partially deteriorated wood pieces. Small blocks made from these samples were boiled in water and 20 μm sections (cross, tangential and radial) were cut on a Reichert Sliding Microtome. The sections were stained in haematoxylin and safranin and mounted in DPX mountant. The wood was converted into chips and macerated for observing individual elements as per Jane (1970). Sections were studied under Leitz Laborlux Microscope connected to a Quantimet 500MC Image Analysis system for anatomical characterisation and identification. Prior to microtoming, the samples were studied for their physical features. The samples were also observed with and without hand lens (10 X) to describe the wood as following Rao and Juneja (1992). The basis of identification of the wood material was a combination of physical features, gross structure and anatomical characters. Photomicrographs were taken from image analysis system (Purkayastha, 1982; and Purkayastha, 1985).

Description of Timber of St George's Reef

The underwater exploration at St George's Reef yielded a large piece of timber lying close to the reef. The maximum length, width and thickness of the timber is 200, 18 and 16 cm respectively. The timber was partially buried in the seabed probably soon after the ship wrecked, hence, it was not severely affected by current and wave actions (Fig. 33). However, one end of the timber is affected by woodborers. The timber is in U shaped and longitudinal rabbets indicate that it could be the part of the keel of the ship. The presence of iron nails show that it was joined to other wooden planks of the ship. As the ship ran aground over the reef and broke into pieces, all the timber, light cargo and other organic material either might have been washed out or disintegrated in the subsequent time. The radiocarbon dating of the timber shows that the timber is 114.3 ± 1.5% years old [measured radiocarbon age T1/2 = 5570 ± 30 years] (Sila Tripati et al., 2003b). The anatomical analysis of wood indicates it to belong to the *Lagerstromia microcarpa syn Lagerstromia lanceolata* species, for which the trade name is 'benteak'.

Description of the Iron Anchor with Wooden Stock of Aguada Bay

The state of the iron anchor of the Aguada bay is good. It appears that the shank and half portion of the wooden stock of the anchor, buried in the seabed, is intact and not at all affected by woodborers. The other half portion, which was above the seabed, is highly affected by woodborers. With the shank at the centre, nine iron nails have been used to join the wooden stock. Out of the nine nails, four nails have been used around the anchor shank and the remaining five nails have been provided on the other part of the stock. Only one nail is visible on the side affected by the woodborers whereas the other four are on the intact portion of the wooden stock. No uniform distance has been maintained between the nails (Fig. 64). A thick coat of yellowish cement mixture has been applied on the surface of the anchor and wooden stock except the iron ring, probably to protect the anchor from the marine environment. Microscopic observation of the coating material shows that the sediment of mica and shell material have been mixed with magnesium and calcium. This mixture becomes hard and does not dissolve in seawater and that could have been used as a protective layer for the anchor. The Infrared spectrum (KBr pellet method) analysis (Fig. 84) of the mixture carried out using a Shimadzu FTIR model 8201PC shows the presence of organic substances on the anchor stock, which could be the deposit of marine organism belonging to the later period. The radiocarbon dating of the wood of the anchor stock

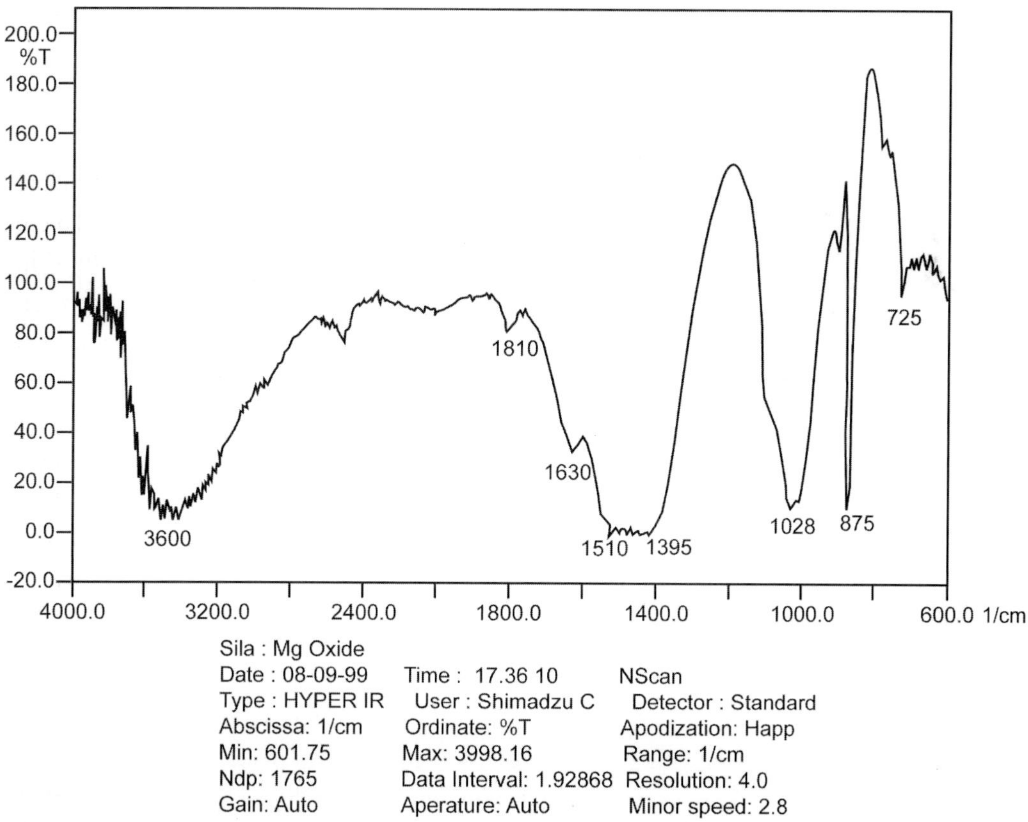

Fig. 84: Infrared spectrum of coating material on the wooden stock of iron anchor

suggests that the wood is 2500 ± 410 years BP; its calibrated age range is 2950-1950 years BP (Sila Tripati et al., 2003a). The anatomical analysis of wood indicates that it belongs to the *Tectona grandis* species for which the trade name is teak.

Results of Timber Analysis

Any Wood Anatomist always looks at the specimen in terms of physical properties and also some of the attributes of internal anatomical structure. Further, the characteristics of any specimen are observed on cross or transverse or end grain side, tangential longitudinal direction and radial longitudinal direction, both at macro level (gross structure) and at microscopic level (minute structure) for description which ultimately will lead to identification. The anatomical study of the wood samples show that the colour of the wooden stock fixed to iron anchor found in Aguada bay is brown, moderately hard, heavy and with medium texture. Its surface is soft rot and soggy in nature. The gross structure of the wood shows that the tree rings of the wood is porous. The ring growths are distinct, conspicuous to naked eye and delimited by large early wood vessels enclosed in parenchymatous tissue. Vessels in the early wood are visible to the eye, which is mostly solitary, oval in outline and gradually becoming smaller towards the latewood. Tyloses are present. Latewood vessels are small, mostly solitary and round to oval in shape. Parenchyma forms thin sheaths round the vessels and it is distinct to the eye in the early wood. Rays are distinct under the lens. The microscopic structure of the sample shows that vessels of wood are up to 285 μm in diameter, transition from early to late wood is gradual; intervascular pitting is alternate, 4-7 μm, vestured; vessel member length is 235-440 μm, tyloses is present and perforation plate is simple. Parenchyma paratracheal delimit the growth of rings. The vasicentric and crystals are absent. Fibres angled in transverse section are not aligned in radial rows, septate, fibre diameter up to 32 μm, lumen diameter up to 25 μm, fibre length is 1027-1383 μm. Inter fibre pits are numerous, confined to the radial walls. The macerated fibres showed extensive damage by softrot fungi as evidenced by softrot cavities (Fig. 85). This suggests that the material was exposed for quite sometime to wet and dry conditions prior to sinking. Rays 1-5 seriate, somewhat heterogenous, ray width is 77 μm and height is 20 plus cells or 688 μm. Silica is present. The diagnosis shows that the wood belongs to the *Tectona grandis*, for which the trade name is 'teak' (Sila Tripati et al., 2005a). The colour of the timber collected from the shipwreck off St George's Reef is brown, moderately hard, heavy and coarse in texture. The observation of the sample under hand lens shows that the timber has darkened surface, which penetrates a few millimetres inside and is slightly soft. A few vessel elements indicate the presence of some non-woody materials. We suspect these may be the remains of marine organisms. The gross structure study shows that the timber

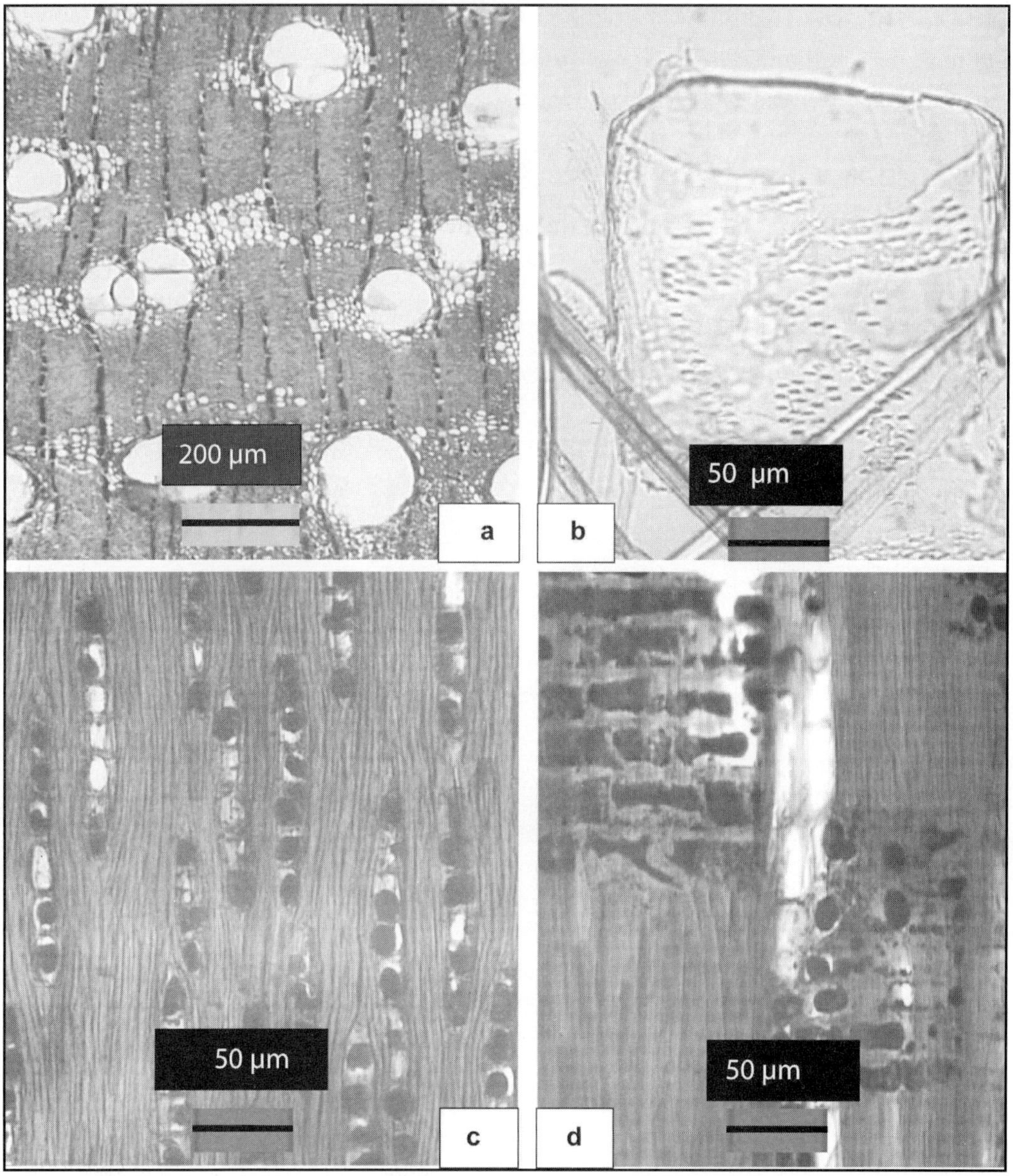

Fig. 85: Anatomical analysis of timber collected from St George's Reef, [*Lagerstroemia* sp.]
a) CS showing vessels, diffuse and aliform parenchyma,
b) Maceration showing vessel element with intervessel pits,
c) TLS showing uniseriate rays,
d) RLS showing procumbent cells with deposits.

is semi-ring porous wood; growth rings are demarcated by large early wood vessels and fine parenchyma bands. Vessels are solitary and in radial multiples of two. The transition from early to late wood is gradual. Parenchyma paratracheal vasicentric to aliform and rays are fine to fine. The microscopic structure study indicated that growth rings are demarcated by large early wood vessels. Vessels are 98-352 µm in diameter. Intervascular pitting alternate, oval, 7-11 µm, vestured, vessel element length 268-550 µm in length, truncated or tailed at one or both ends. Perforation plate is simple Parenchyma paratracheal, aliform to aliform confluent and forming short bands. Moreover, crystals are present and fibres are rounded to angular, more or less radially aligned in cross section. The fibre diameter is up to 32 µm and lumen diameter is up to 25 µm, fibre length is 1003 -1479 µm, septate with several septa in each fibre. Interfibre pits are minute, simple rather sparse, confined to the radial walls. Rays are uniseriate rarely biseriate homogenous, ray width is up to 47 µm and height 15 plus cells or 340 638 µm (Fig. 86). The diagnosis shows that the wood belongs to the *Lagerstromia sp.* probably *Lagerstromia lanceolata* for which the trade name is 'benteak'.

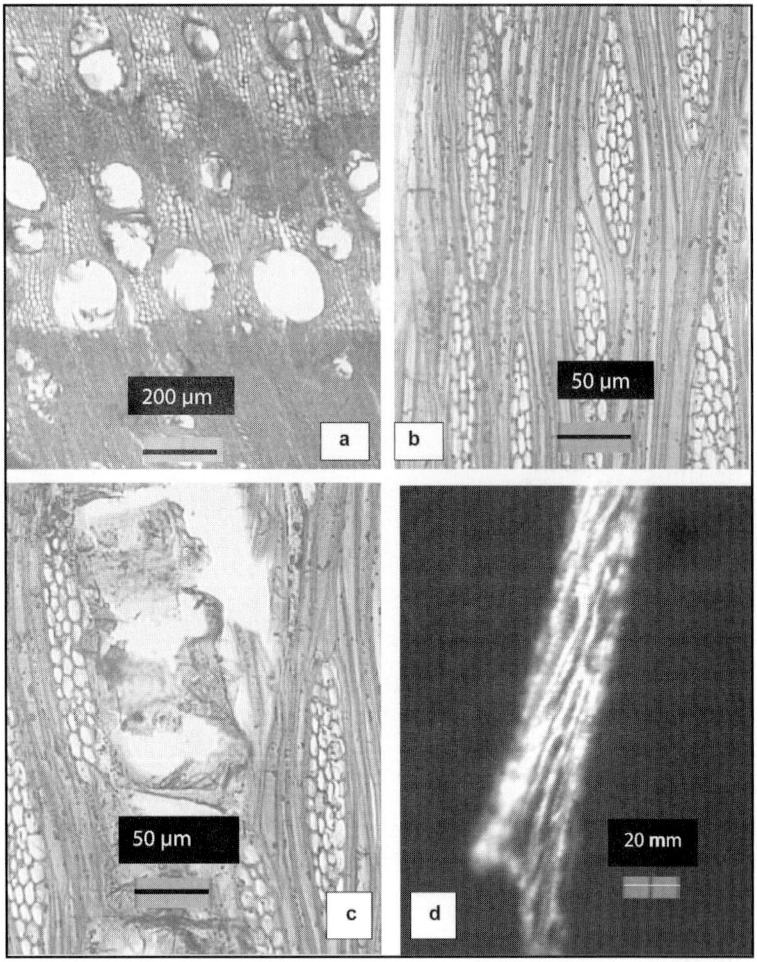

Fig. 86: Anatomical analysis of timber of the wooden stock of the anchor recovered from Aguada Bay, [*Tectona grandis. a,* CS showing ring porosity, parenchyma distribution. *b, c.* TLS showing ray morphology, intervessel pits and septate fibres *d.* Fibre in polarized light showing soft-rot cavities due to fungal attack].

Discussion

Numerous ships had been built both in India and Lisbon for the maritime trade with India but no records are available as to how long those were used their, abandonment and other details. However, one of the reasons, why Afonso de Albuquerque, conquered Goa in 1510 AD, was to capture the shipbuilding industry of Goa (Mathew, 1988). During the Portuguese period the shipbuilding industry of Goa reached to its zenith. There were three dockyards in Goa between the city of Panaji and on the riverbank of Mandovi (De Kloguen, 1995). A number of ships were built in the Goa shipyard among them the important ones were *Nau Cinco Chagas, Galeão Bom Jesus, Madre de Deus,* and *Nau São João Baptista.* Before coming to India Portuguese used oak and pine timber for construction of ships. These ships were frequently affected by woodborers, crumbling due to weak timber and maintenance cost of the ships was relatively high compared to their longevity. However, the native sailors of India used teak, which was the best timber ever known for shipbuilding industry in India. Subsequently, Portuguese understood the superiority of teakwood for construction of ships and used teak extensively in the shipbuilding industry. For the purpose they established shipyard and dockyard in Goa for construction of ships. Teak was available in large quantity in the western coast of India particularly in Kerala, Karnataka, Maharashtra and Goa. The Portuguese as well as Indian boatbuilders used readily available teak in construction of their ships and boats. During the initial stages the *Basel Mission Company* used bullock carts for transporting their products in the regional area. In course of time the demand for the products increased and as a result the company used native crafts on hire for the transportation of products throughout India (Raghaviah, 1990). The hiring of native crafts was cheaper and huge quantities of products were transported without hindrance. Tiles and other products were transported to Bombay (Mumbai) by native crafts and from Bombay those were exported to the British East Africa, Aden, Australia, Basra, Borneo and Sumatra (Hofmann, 1913). The company also operated in other parts of the world, namely Ghana, Borneo and China but their industrial base remained the strongest on the western coast of India because of low investment costs and ready availability of manpower. The *Basel Mission Company* used to engage wooden sailing native crafts (with and without cabins) for transportation of the finished products, which is evident from the photographs of 1910 AD (Raghaviah, 1990). Probably those crafts could be either *Kotia* or *Vana* type of crafts. Their sizes ranged between 8 to 10 m in length and had the capacity to carry 30 to 50 tons. These crafts could be sailed very close to the shore and the draught would be 2 to 2.5 m. The timber, which was found in St George's Reef, may have had similar dimensions to those mentioned above. These vessels were predominant on the Western Indian Ocean countries and were meant for transportation of cargo and local products. The timber found in St George's Reef belongs to a craft, which could be smaller in size and locally built, because during the 19th century the size of the European ships was much larger and bigger especially when compared with the native crafts of India. Hence, it appears that the timber belonging to the craft could not have been of foreign origin and the probability of an Indian origin cannot be ignored.

Though it is difficult to state the provenance of the timber found in the wooden stock of Aguada bay and from the single timber from St George's Reef, it is strongly assumed that the timbers could be of Indian origin. This is evident from the fact that *Lagerstromia microcarpa* being a species of Western Ghats and known for its use in boatbuilding industry might have been utilized for the purpose from any of the localities along the west coast of India. Additionally, the boatbuilding and shipbuilding industry of India was known to the mariners of the other overseas countries since ancient times. Portuguese had established shipbuilding centres and dockyards along the west coast of India namely at Old Goa, Thane and Bassein because of availability of good teak, which was cheaper, more durable than other timber and readily available skilled craftsmen (D'Silva, 1997). Therefore, the necessity for making wooden stock to get timber from other countries did not arise. Some portion of the wooden stock is affected by woodborers due to exposure to marine environment, which could have been caused subsequently. The timber is not decomposed in spite of lying underwater since centuries together exposing to the marine environment and this could be due to the quality of teak wood. There are instances of Indian built ships namely *Sydney Cove* and *Valetta* wrecked off Australia, which were built in Calcutta (Kolkata). The timbers used in the construction of the *Sydney Cove* were teak (*Tectona grandis*), *sisoo* (*Dalbergia sisoo*) and Indian rosewood. The anchor stock of *Sydney Cove* was also made from teak (Nash, 2002b). The timbers used in the construction of the *Valetta* have been identified as teak; the report states that there were other local (Indian) timber and material were used in the construction but it is unable to identify them (May, 1986). The other Portuguese ship *Santo Antõnio de Tanna* (1697), which was built at Thane, near Mumbai and wrecked off Mombasa was also made of Indian teak wood, which was found in the forest of Bassein near Mumbai. Archival records related to the building of the ship kept at Goa State Archives, Panaji and the writings of Portuguese naval experts, navigators and sailors of the that time give other details about *Santo Antõnio de Tanna* (Piercy, 1977; and Blot, 1991; Blot and Blot, 2001). Similarly, teak was used for construction *of HMS Trincomalee* which was built in Bombay in 1817. *HMS Trincomalee* frigate served in the Royal Navy of the East India Company till 1897; afterwards it was refitted and renamed as *Foudroyant* subsequently sold (Lambert, 2002).

Conclusion

Teak is one of the best timbers used for various purposes. The shipwrights industry also used teak in the shipbuilding industry for its noble qualities. The anatomical analysis indicates that *Lagerstromia microcarpa* and *Tectona grandis* grow in the western ghats of India. The Indian as well Portuguese shipwrights used both *Lagerstromia microcarpa* and *Tectona grandis* species in construction of ships. The timber off St George's Reef corroborates this. Portuguese understood the qualities of teak, they not only used it extensively in shipbuilding industries but also for various other purposes such as furniture, house and church constructions. The East India Company also built several ships both on the east and west coast of India for the Royal Navy and overseas trade and commerce. Use of Indian teak for construction of ships by the Portuguese, British and Indian boatbuilders has been proved from the shipwreck findings both in Indian and foreign waters. The use of Indian teak by the Portuguese for making anchors is confirmed from the finding of wooden stock of iron anchor of Aguada bay. The use of teak for other purposes in the shipbuilding industry and their provenance could be known from the findings of shipwreck explorations in India in the coming years.

11

Analysis of Timber and Coating Material on an Iron Anchor Recovered off Aguada Bay, Goa

Timber played an important role in the shipbuilding industry since ancient times, particularly for the construction of seaworthy boats and ships including stone, lead and iron anchors. For instance timber was used as flukes for stone anchors, shank for lead stock anchors and also stock for iron anchors. Although lead stock anchor with wooden shank has so far not been recovered from Indian waters, wooden flukes in stone anchors and wooden stock in iron anchors were reported from offshore regions of Maharashtra (Gaur et al., 2007a) and Goa (Sila Tripati et al., 2005a) in the west coast of India. Indian literature namely *Yuktikalpataru* vividly mentions about ancient shipbuilding activities i.e; various parts of the ships, types of ships used for trade, warfare, pleasure trips and kind of timbers used in their construction but hardly any reference is made about the use of timber in stone and iron anchors. In the *Astadhyayi*, Panini (5th century BC) mentions the use of large variety of timbers namely *amra* (*Mangifera indica*), *khadira khair* (*Acacia catechu*), *simsapa* / blackwood (*Dalbergia sissoo*) (*Delbergia Sisoo*), salmali / silk and cotton tree (*Bombax malabaricum*) (Agrawal, 1953). The *Periplus of the Erythraean Sea (Periplus Maris Erythraei)* enumerates teak wood, black wood and ebony as the important wood varieties exported from Barygaza to the ports of Apologos (Apologues) and Ommana on the Persian Gulf coast (Warmington, 1974). During the rule of king Nebuchadnezzar (604-562 BC), India is said to have exported timber to Mesopotamia for shipbuilding and also his palace is said to be built with teak from India (Rao, 1970). Hourani (1995) states that Indian teak is most precious and valuable due to its durability, elasticity, strength, non-yielding to cracks, splits and shrinks.

Iron Anchors in Goa waters

During the Portuguese period, Aguada bay of Goa was used as a sheltering and refuelling stop for the ships en-route to Old Goa, as well as Portugal. Portuguese records state that while crossing the sandbar [sandbar is a somewhat linear landform within or extending into a body of water, typically composed of sand, silt or small pebbles] in the River Mandovi, a number of ships have been wrecked. This sandbar was referred in the Portuguese Hydrographic charts as *Barra de Agoada.* During 1977-82, the sandbar was dredged leading to the recovery of four iron anchors of different types and a number of timber pieces. These anchors are presently housed in the Goa State Museum, Panaji (Sila Tripati et al., 2003a). In 1999, Florenco Marques (owner of a local fishing boat) found an Admiralty Long Shanked iron anchor off Aguada bay in the '*satbahu*' in 14 m water depth and reported the same to NIO, Goa. Another anchor of similar type, but without the wooden stock had earlier been found from a shipwreck site at Sunchi Reef (Sila Tripati et al., 2006a). One more prototype of Admiralty Long Shanked iron anchor with wooden stock was recovered from 11 m water depth (15^0 26 568 N and 73^0 46 950 E) in the vicinity of lighthouse of Aguada bay in January 2008 (Fig. 64). Anatomical analysis of the wooden stock was carried out in order to identify the species of the timber and their probable provenance. In addition, Scanning Electron Microscope (SEM) and Energy Dispersive Spectrum (EDS) analyses of fine powdered layer applied on the anchor surface has been carried out in order to understand its elemental composition, purpose of its application and to tentatively date the anchor in relation with the maritime history of Goa. Further this study would provide use of timber other than teak in shipbuilding industry.

Description of the Iron Anchor

The anchor collected from Aguada bay is intact, with a length of 3.30 m and flukes measuring 1.35 m. The diameter of the shank varies from 45 to 47 cm and iron ring is 2.5 cm thick and 55 cm diameter provided for tying cable. The length of the wooden stock is 4.37 m and made of two timbers joined together with nine nails (pins) (Fig. 66). The thickness of each timber varies from 20 to 30 cm. Incidentally length of stock is much more than the anchor. A well preserved uniform thin layer of protective coating has been observed on the surface of the flukes, shank and nails of the wooden stock. The upper stock is covered with huge deposits of shell colony (bivalves) (Fig. 87) in addition to the deterioration by the

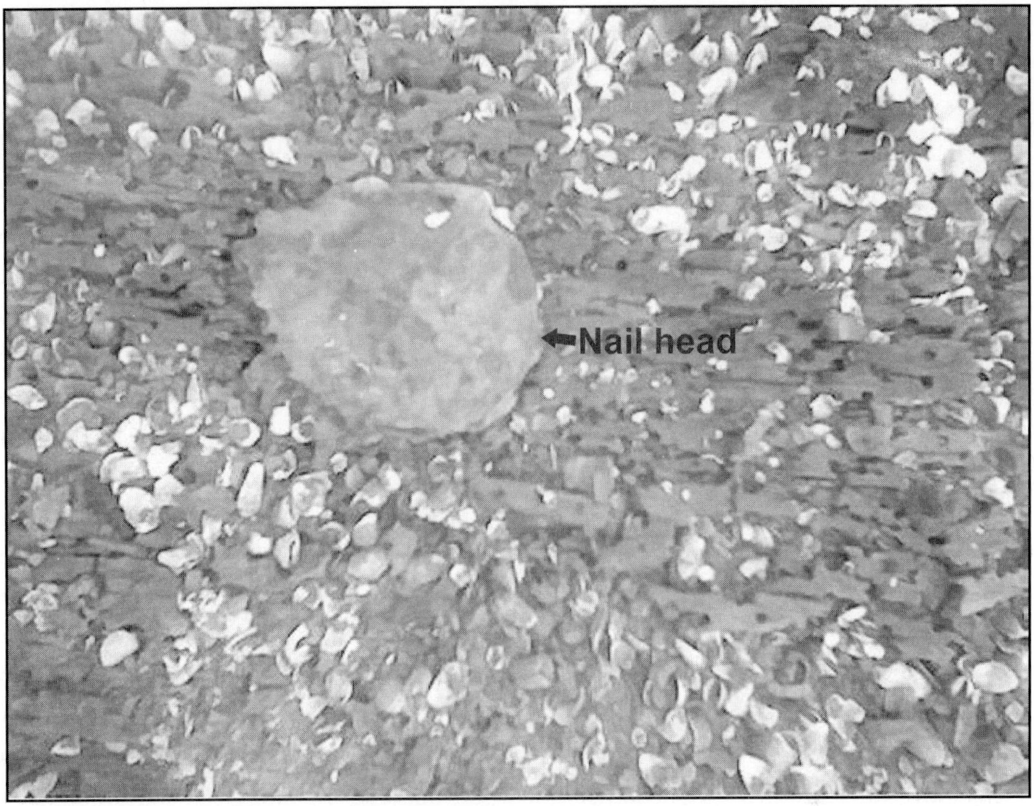

Fig. 87: Nail head and deposition of bivalves on the wooden stock of the anchor

woodborers at a few places. Similarly, growth of oysters was noticed on the flukes of the anchor. However, the lower timber, buried in sediment, is harder and has not decomposed like the upper timber. Nine iron nails are used to join the upper and lower timber of the stock and keeping the shank at the centre of the stock. Nails are placed at varied distance and remained intact with the coating layer. No iron bands or strips have been used for further fastening of the wooden stock, as found in the anchors of late 18th and early 19th centuries (Curryer, 1999; and Upham, 1983).

Anatomical Analysis

In order to identify the species of the timber, its quality, state, suitability for making as a stock, samples were collected from both the timber pieces of the stock. The material used for anatomical study consists of small un-carbonized and partly deteriorated wood pieces. The samples were observed with and without hand lens (10X) to describe the wood, following the methods of Rao and Juneja (1992). Small blocks were made from these samples and boiled in water and 20 µm sections (cross, tangential and radial) were cut using Reichert sliding microtome. The sections are stained in haematoxylin, safranin and mounted with DPX mountant. Sections were studied under a Leitz Laborlux microscope connected to a Quantimet 500 MC Image Analysis system for anatomical characterization and identification. Part of the wood was then converted into chips and macerated for observing individual elements (Jane, 1970). Photomicrographs were taken by an Image analysis system. Specific gravity was determined using the formulae: mass of the sample at test / oven dry weight of the sample.

SEM and EDS Analysis

Further, to determine the elemental composition of the coating material, SEM and EDS studies were carried out at spots on these flaky coating specimens. The specimens were cleaned in ultrasonic bath to remove the loose material deposited from the water column. Then these flakes were placed on a carbon conductive tape stuck on a nylon stub. To minimize the effect of depositional material, the specimen peeled from the anchor was mounted on the stub in such a way that the lower surface of the specimen to be analyzed faced upwards. Afterwards the specimens were sputter coated with about 20 nm thick gold coating using a gold sputter coater. Coated specimen images were then analyzed using SEM (model JSM 5600) with an EDS attachment (model JOEL 5800 LV). Analysis of timber was carried out for identification of wood material including the combination of physical features, gross structure and anatomical characters as per IAWA (*IAWA Bulletin*, 1989) list of microscopic features for hardwood identification.

Upper Timber Stock

The upper timber of the stock is dark in colour. Its outer side is soggy, soft, peeling off easily, while the inner side is non-soggy, hard, heavy (sp.gr.0.875, oven - dry) shallowly interlocked, grained and even textured. In case of gross structure of the timber, growth rings are distinct. Wood is diffused porous. Vessels are distinct, under hand lens, mostly solitary, open and vessel lines are distinct. Parenchyma is paratracheal, aliform, aliform confluent forming zigzag arrangement due to the vessel distribution. Rays are fine and closely spaced. With reference to the minute anatomy of the timber, vessels are found in radial multiples of 2-3 (44%), singles (28%) and clusters (28%). Solitary vessels are oval in outline, infrequently filled with yellowish deposits. Vessels are large, with tangential diameter in the range of 141-213 µm: av= 181 µm. Vessel frequency varied from 4 to 5/mm2 and their element length was found in the range of 271-410 µm: av= 330 µm) with simple perforations. Inter vessel pits are alternate, round to angular in shape and medium to large in size i.e., 7-11 µm: av= 9 µm and vestured. The vessel ray pits are similar to inter vessel pits. Fibres are filled with dark coloured deposits, septate, thin to thick walled. Their length ranged from 837 µm -1311 µm (average 1098 µm). Fibres from the surface area of the sample showed soft rot cavities. Axial parenchyma is paratracheal, aliform, aliform confluent, joining 2-4 vessels forming zigzag pattern; also diffuse, consisting of 4 cells per strand without crystals but full of dark deposits. Rays are predominantly uni-seriate, 8-12/mm, homocellular to weakly heterocellular composed of procumbent with one row of upright/ square marginal cells. Some rays are partly bi-seriate. Ray height varied from 203-473 µm, with dark coloured deposits (Fig. 88). Prismatic crystals are observed in procumbent, upright and square cells. Heavy deposits obscured the observation of crystals under compound

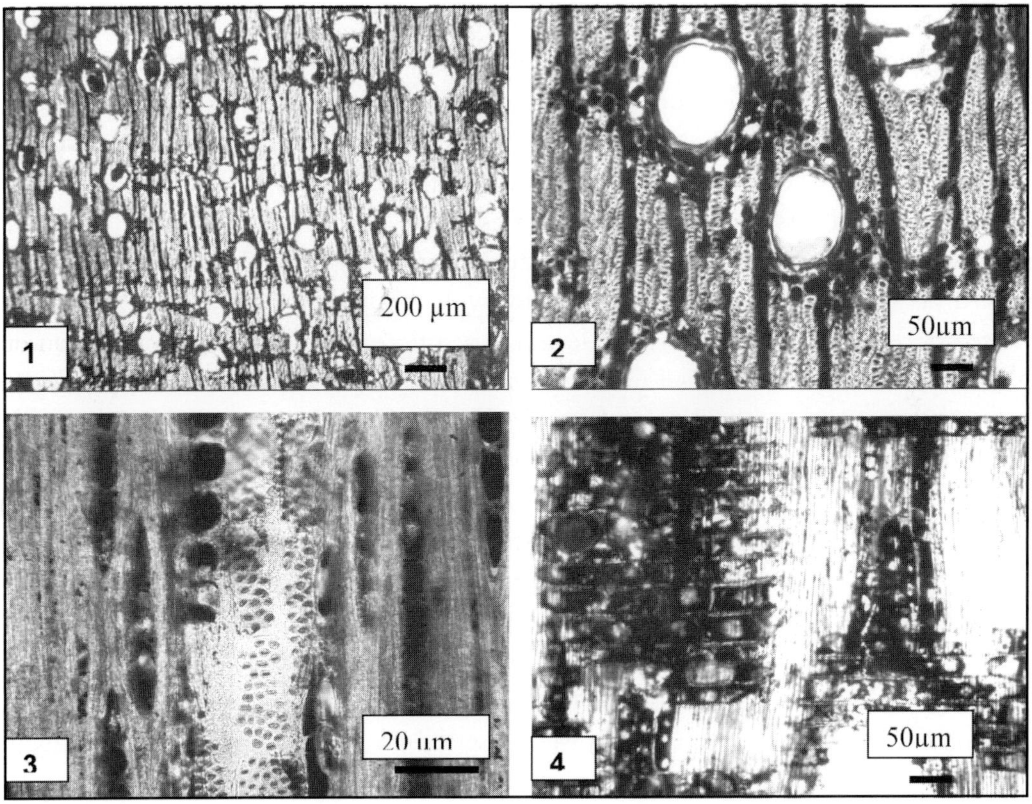

Fig. 88: Anatomical analysis of timber from the upper wooden stock (*Terminalia* spp.)

Fig. (1) & (2) CS showing diffuse porous wood with aliform and aliform confluent parenchyma. Parenchyma contains dark colored deposits.

Fig. (3) TLS showing uniseriate rays with dark colored deposits and inter-vessel pits.

Fig. (4) RLS showing weakly heterocellular rays with dark colored deposits.

microscope and under polarized light there is some loss of bi-refringency although crystals are visible. The card key features of upper timber as per IAWA, 1989 are as follows: 1, 5, 13, 22, 23, 26, 27, 29, 30, 42, 46, 52, 56, 60, 65, 69, 72, 79, 83, 92, 96, 106, 115, 137, 138, 189, 192, 195. The diagnosis of the timber has been carried out based on the characters mentioned above and the sample has been identified as *Terminalia* spp. belonging to Combretaceae.

Lower Timber Stock

The general feature of the lower wood is dark brownish in colour with outer side soggy, soft, peeling off easily, inner side non-soggy, hard and heavy (sp. gr.0.88, oven-dry), deeply interlocked grained and even textured. As far as the gross structure of the lower timber is concerned growth rings are distinct to indistinct. Wood is diffused porous. Vessels are visible under hand lens, moderately numerous, solitary, predominantly in radial multiples of 2-4, open, vessel lines are short, distinct under hand lens and irregularly oriented because of the inter-locked grain. Parenchyma is not visible because of the dark background. Rays are fine and short because of the distorted orientation of fibres. The minute anatomical study shows that vessels are in radial multiples of 2-3 rarely up to 4 or more and occur mostly as clusters. Solitary vessels are oval to angular in shape and its tangential diameter varied between 62 and 104 µm: av=88 µm) with thin walled tyloses. Vessel frequency varied from 13-20/mm2 and average length varied from 214-612 µm: av= 430 µm with simple perforations. Inter vessel pits are alternate, polygonal and large, 10-14 µm: av 12 µm in size, non-vestured. The vessel ray pits have highly reduced border to simple pits, horizontal, 11-21 µm: av 15 µm in size. Fibres are with simple to minutely bordered pits, septate, thin to thick walled. Fibre length varied in the range 1097 -1675 µm: av 1446. Axial parenchyma is vasicentric and consisting of 3-4 cells per strand. Crystals are not observed. Rays are 1-3 seriate, predominantly 2-3 seriate, 33-56 µm: av=41 µm wide, 6-8/ mm heterocellular and composed of procumbent and one row of upright/square marginal cells having colored deposits. Inclusions are not observed. Ray height varied between 213 and 453 µm: av=323 µm or 20 cells high). Crystals of variable types were observed in procumbent or upright cells. Oil cells were observed predominantly in axial parenchyma (95 µm in length and of 57 µm in width in RLS) (Fig. 89).

The card key features of the lower timber as per IAWA, 1989 are: 1(v), 2, 5, 12(v), 13, 22, 27, 32, 41, 47, 53, 56, 61, 65, 69, 72, 79, 92, 97, 106, 115, 125, 152, 189, 192, 195. Diagnosis of the timber has been carried out based on the characters mentioned above and the sample has been identified as *Phoebe* spp. belonging to Lauraceae.

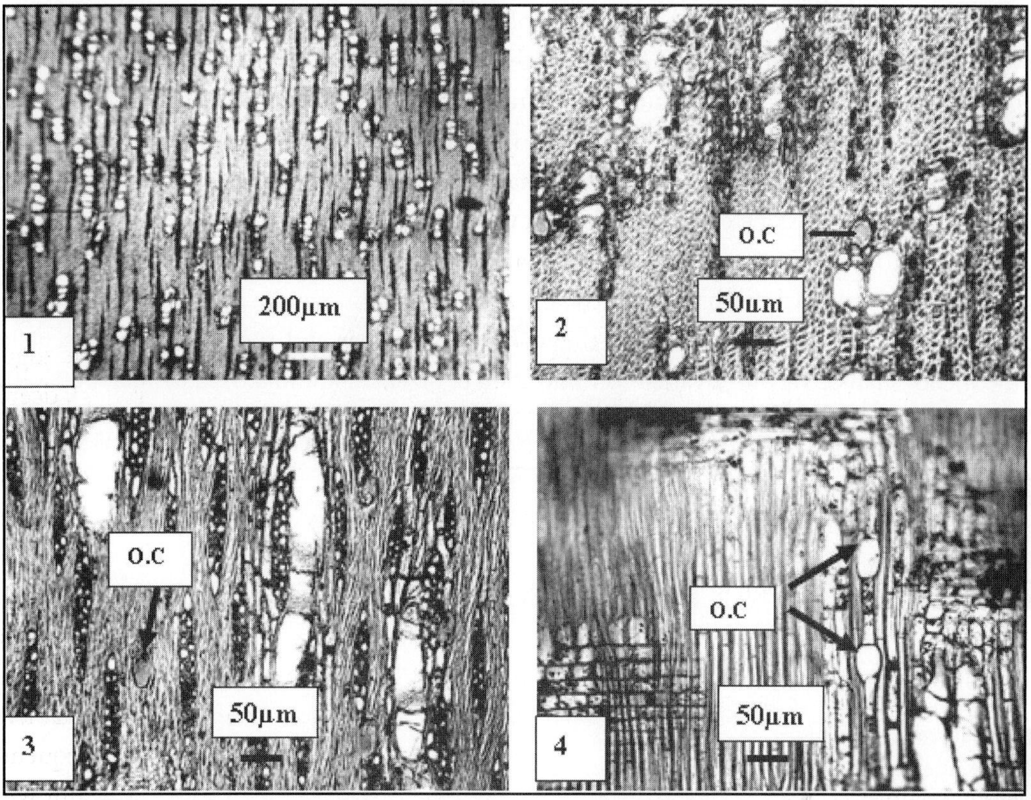

O.C - Oil Cell (see arrows)

Fig. 89: Anatomical analysis of timber from the lower wooden stock (*Phoebe* spp.)

Fig. (1) & (2) CS showing diffuse porous wood with indistinct to distinct growth rings.
Fig. (3) TLS showing multiseriate heterocellular rays and vessel members with inter-vessel pits and axial parenchyma with oil cells.
Fig. (4) RLS showing prominent oil cells in axial parenchyma and rays with procumbent and upright cells.

Note: In view of the fragile nature of the wood, card key features from 164 to 188 and 196 onwards have not been taken into consideration for both the timber samples.

SEM - EDS Analysis of Coating Material

From the coating layer on the anchor, several thin flakes from clean areas have been collected from different spots from wooden stock and anchor shank for SEM and EDS elemental studies and results of representative flakes have been presented in Table. 8. The results indicate that silicon oxide (SiO_2) is predominate and its content varies ~61-93 % in the coat on wooden stock while it ranges 43-85% in the coat on anchor shank. Higher SiO_2

content has been observed at the expense of decreased alumina (Al_2O_3) in samples from wooden stock and anchor shank (Table - 8: Flake 1-9 and Flake 1-17). FeO content in the coat from wooden stock is low and ranges from 2 to 5% when compared to 11 to 22% in the coat of anchor shank. Na_2O is high (6%) in the coat on the stock when compared to negligible (<0.5%) content in the anchor shank coat. Interestingly, MgO content shows a reverse trend with its low contents (<1%) in the coat on wooden stock compared to higher values (up to 10%) in the coat of anchor shank. Similarly, FeO is less (2.65-4.65%) in the coat on wooden stock with high (11.69-21.67%) FeO contents in the coat from anchor shank. K_2O and TiO were absent in the wooden stock coat and conspicuously present in the anchor shank coat with a varied concentration of 0.2-2.94% and 0.81-2.07% respectively.

Element	Flake1-8	Flake1-9	Flake1-14	Flake1-15	Flake 1-16	Flake 1-17
	Coating layer on wooden stock	Coating layer on nail heads and iron shank				
Na2O	6.21	—	6.05	—	0.44	—
MgO	0.71	0.97	0.43	10.34	6.58	0.72
Al2O3	21.74	2.31	21.76	21.66	28.05	1.68
SiO2	61.46	92.87	61.37	43.45	46.99	85.28
K2O	—	—	—	0.22	2.94	0.29
CaO	7.22	0.52	5.74	0.59	0.27	0.33
TiO	—	—	—	2.07	0.81	—
MnO	—	—	—	—	0.36	—
FeO	2.65	3.32	4.65	21.67	13.55	11.69

Table 7. Weight percent of elemental oxides of thin coating material from the wooden stock and Iron Anchor.

Discussion

The finding of the anchor with wooden stock off Aguada bay in the recent past further confirms the use of good quality timber in the shipbuilding industry during the maritime history of Goa and Portugal. The Indian Ocean region is of tropical nature; hence the attack of woodborers is comparatively more than in the other seas. Probably this might be the reason that the application of a coat of fine-grinded material on the surface of the anchor as well as on the wooden stock was made perhaps to protect them solely from the marine environment. It is difficult to diagnose the individual mineral grains in the coating layer under microscope, as it constitutes very fine powder. Coating layer on the anchor is made of fine powdered material presumably derived from either beach or river sorted sands. Beach sand is moved by near shore wave generated currents. Highly variable wind and wave regime in the coast determine the fate and final composition of beach sands, which accrete and erode beaches. Beach sand composition mainly depends on regional geology, more particularly bedrock mineral assemblages in the hinterland and mostly contains mineral sands, biogenic sands and the mixtures of these. SEM and EDS analysis of the thin flakes (Fig. 90) shows that the composition of this layer is nearly homogenous indicating the use of well sorted sands for protective coating on the anchor. The composition of coating material is analogous to beach sands than river sands. River sands are generally more variable and divergent in composition than the beach sands due to high energy waves swash on the beach front. Most beach sands, however, are composed of grains of the mineral quartz due to its higher relative density. In addition to the predominant quartz, beach and river sands contain other lithics such as feldspar, mica, hornblende, ilmenite, magnetite, garnet and also a variety of trace minerals such as iron/titanium oxides, zircon, monazite and calcite. Substantial TiO_2 (0.81-2.07%) found in the iron anchor coat may be due to the presence of garnet as garnet is commonly present in lithics derived from hinterland. Titanium may be present in trivalent state in garnet along with Fe and Al. Based on the variations in the contents of oxides of Na, Mg, K, Ca, Ti and Fe in the coats of wooden stock and anchor, it appears that source material from two different beach sands may be used for these coatings.

It is generally believed that coating materials were not applied on iron anchors and wooden stocks. Conversely, a uniform layer would not have existed on the surface of the anchor as well as either sides of the wooden stock. This type of anchor would not be ground or roll on the seabed like the stone or lead stock anchors. When anchored, either one fluke or one side of the stock would remain above the seabed. If coats were a deposit from water column, it would not have been present on the buried portion of the anchor. This fact also is corroborated in the compositional differences in the coats from wooden stock and anchor. It could be possible that such type of coats, constituting well sorted sands from beaches, were applied on the anchors to maximise its endurance in tropical waters, where the nutrient rich water support faster growth of wood borers which, in turn, could deteriorate the stocks. It is known from the records as well as local mariners that a mixture made of lime and other ingredients applied on the ships, coir, etc. to protect them from woodborers. Another effective deterrent against woodborers is cashew nut oil, which is also very popular among mariners. In the earlier study (Sila Tripati et al., 2005a), it was observed that the wooden stock of the iron anchor was made of teak (*Tectona grandis*) and benteak (*Lagersrtoemia sp.*) indicating a preference for tropical timbers for maritime trade. The present study indicated that iron anchor with wooden stock is made of *Terminalia* spp. and

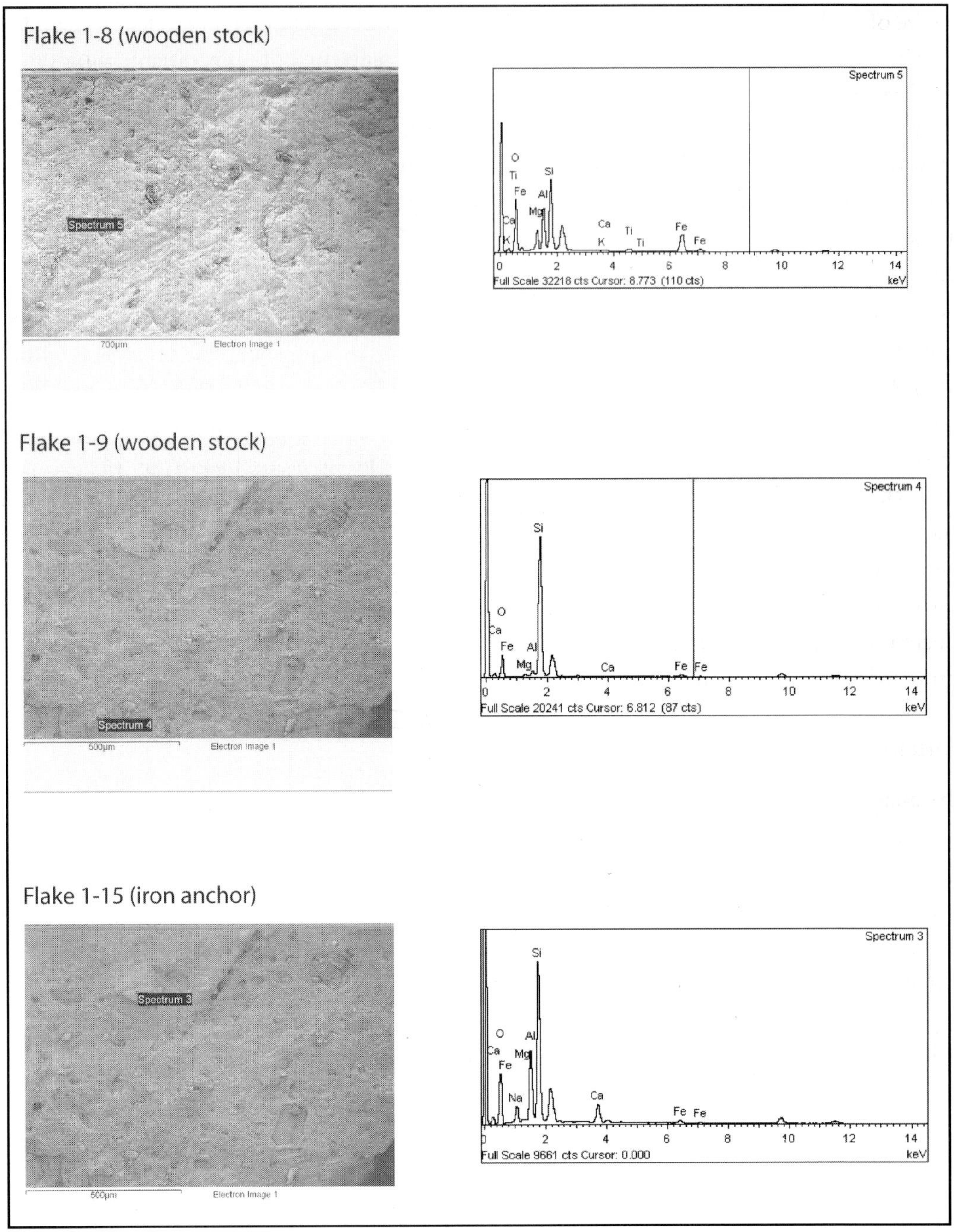

Fig. 90: SEM-EDS analysis of representative parts of the coating material from the wooden stock and anchor shank

Phoebe spp. ranged from 250 and 70 respectively the former having distribution in Tropics of the world and the latter in Indo-Malayan region, Pacific Islands, Tropical America and the West Indies (Annos, 1969; and Anon, 1976). Limited studies made on wood anatomy of these two genera (Stern, 1954; Purkayastha and Lal, 1972; Vliet van, 1979; Negi and Purkayastha, 1985) indicated that identification of the timber at species level was difficult. In view of this, it is not possible to comment on which of the species has been used for making wooden stock of the anchor. Furthermore, the distribution of these two genera across many countries makes it very difficult to pinpoint the exact provenance from where the timber has come. The then sea route, including the timber trade by Portuguese community may throw some light on the source of timber used for this purposes. The heavy weight and quality of these two timbers might be one of the criteria for their selection as wooden stock of the anchor. It is interesting to note that *Terminalia* and preservative treated *Phoebe* woods were known for their use in shipbuilding, oars, shafts, masts and hatch covers (Purkayastha and Lal, 1972; Vliet van, 1979; Negi and Purkayastha, 1985; Sharma et al., 1974). The mariners might have outsourced the raw material from any en-route countries in their voyages. *Phoebe* has not been tested for durability studies under marine condition, whereas limited studies have been made on a few species of *Terminalia* at selected localities. As indicated by us that both the timbers are heavy, hence might have entranced the shipbuilders for its use as wooden stock of iron anchor.

Conclusion

The anatomical study shows that not only teak but also other timber such as *Terminalia* and *Phoebe* spp. were used for shipbuilding by the Portuguese to have better endurance of timbers in tropical marine conditions. Moreover, Europeans used timbers from India and other countries for the construction of bigger sized seagoing vessels and also for stock of the iron anchors. These timber species are not found in India. This is the first report from Indian waters about recovery of timber of non teak type wooden stock. The absence of iron strips, which are common among the anchors made in the 18th-19th centuries, probably indicate that this stock is contemporary of the 16th-17th centuries. This coating material on anchor and stock has not been previously reported. It is presumably applied to have better endurance and protection in the tropical waters. SEM-EDS results revealed that the composition of fine grinded material applied on stock and shank are from two different sources of beach sands.

Bibliography

Agius, D. A., 2005. *Seafaring in the Arabian Gulf and Oman: People of the Dhow.* Kegen Paul Limited, London.

Agrawal, V. S., 1953. *India as known to Panini.* University of Lucknow, Lucknow.

Ahmed, A., 1991. *Indo-Portuguese trade in the Seventeenth Century 1600-1663.* Gian Publishing House, New Delhi.

Alves, J. F., Blot, M., Caleja, P., Carvalho, P., Henriques, R., and Rodrigues, P., 2007. A preliminary analysis of shipwreck data concerning Portugal and India based on written and Archaeological sources. In *India and the Eastern Sea,* (Ed.), A. Tripathi, Agam Kala Prakashan, New Delhi: 97-119.

Alves, J. F., Castro, F. V., Rodrigues, P., Gracia, Catarina., and Aleluia, M., 1998. Archaeology of a Shipwreck. In *Nossa Senhora dos Mārtires, The last Voyage,* (Ed.), S.L. Afonso, Pavilhao de Portugal Expo' 1998 Lisboa, Verbo: 183-215.

Ambre, N. V., 1991. Topography and surface sediment distribution around Amee shoal, Mormugoa. In *Recent Advances in Marine Archaeology,* (Ed.), S. R. Rao, NIO, Goa: 113 -115.

Anon, 1969. W*ealth of India (Raw materials),* Vol. 8, SIR, New Delhi.

Anon, 1976. Wealth *of India (Raw materials),* Vol. 10, CSIR, New Delhi.

Anon., 1981. *Power diffraction file inorganic phase – selected power diffraction data for minerals,* JCPDS, USA, Vol. 5, pp. 627.

Arasaratnam, S., 1994. *Maritime India in the Seventeenth Century,* Oxford University Press, New Delhi.

Arroyo, H., 1975. Contribution for the reconstruction of ancient stone anchors. *Cahiers darcheologie Subaquatique*, Vol. 4: 147.

Athiyaman, N., 2001. Stone anchors from east coast of India a precursor for marine archaeology of Tamil Nadu coast. *PIHC, 61st (Millennium) Session*, 2001: 1293-1310.

Barata, J. G. P., 1989. *Estudos de Arqueologia Naval.* Vol. II, Lisbon.

Barbosa, Duarte., 1967. *The Book of Duarte Barbosa*, Vol. 1, Kraus Reprint, Nendeln.

Barbosa, D., 1970. *A Description of the Coast of East Africa and Malabar.* (Reprint), London.

Barnett, L. D., 1982. Inscriptions at Narendra. *Epigraphia Indica,* Vol. 13: 298-326.

Bass, G. F., 1972. *A History of Seafaring based on Underwater Archaeology*, Thames & Hudson, London.

Bass, G. F., 1996. *Shipwrecks in the Bodrum Museum of underwater archaeology.* Museum of Underwater Archaeology Bodrum, Bodrum.

Bhu, H., Sarkar, A., Purohit, R., Banerjee, A., 2006. Characterization of fluid involved in ultramafic rocks along the Rakhabdev Lineament from southern Rajasthan, northwest India. *Current Science*, Vol. 91 (9): 1251-1256.

Blackmore, J., 2002. *Manifest Perdition: Shipwreck narrative and the Disruption of Empire.* University of Minnesota, Minneapolis.

Bliss, E. M., (Ed.) 1890. *Encyclopaedia of Missions*, Vol. 1. New York.

Blot, J. Y., 1991. The Mombassa Wreck Excavation. *INA Newsletter,* Vol. 18 (2): 6-8. Bowden, J. K., (Ed.) 1971. *Verbal notes and sketches for marine engineer officers: A manual of Marine Steam Engineering Practice.* Vol. 1, James Munro and Company Ltd, Glasgow.

Blot, J. Y., and Pinheiro Blot, M. L., 2001. Archives and Nautical Archaeology: Late 17th century Portuguese shipbuilding in India and the *fragata* '*Santo António de Tanna*'. In *Fourth Centenary Volume of the Goa Archives (1595-1995)*, (Ed.), S. K. Mahmai, Directorate of Archives and Archaeology, Panaji, Goa: 104-124.

Bowen, R. L., 1956. Boats of the Indus Civilization, *The Mariner's Mirror*, Vol. 42 (4): 279-290.

Bowrey, T., 1905. *A Geographical account of Countries round the Bay of Bengal, 1669 to 1679.* Asian Educational Service, New Delhi.

Boxer, C. R., 1957. *An Introduction to the História Trãgico Marítima.* Lisbon.

Boxer, C. R., 1959. *The Tragic History of the Sea.* Haklyat Society, Cambridge University Press, Cambridge.

Boxer, C. R., 1960. The Carreira Da India 1650-1750. *The Mariner's Mirror*, Vol. 46 (1): 35-54.

Boxer, C. R., 1969. *Portuguese Seaborne Empire 1415-1825*. Hutchinson & Co, London.

Boxer, C. R., 1980. *Portuguese India in the Mid-Seventeenth Century*. Oxford University Press, New Delhi.

Brown, R. R., 1984. Comment on the 15th century Hand-Gun ex Association. *IJNA*, Vol. 13 (3): 243.

Bragança Pereira, A. B., 1937. *Arquivo Portugues Oriental.* Vol. II, Bastora.

Brophy, G. P., and Nash, J. T., 1968. Compositional, infrared and X-ray analysis of fossil bone. *The American Mineralogist*, Vol. 53 (3-4): 445-454.

Casson, L., 1989. *The Periplus Maris Erythraei*. Princeton University Press, New Jersey.

Castro, F. V., 2001. The remains of a Portuguese Indianman at the mouth of the Tagus, Lisbon, Portugal (Nossa Senhora dos Mãrtires, 1606?). In *Proceedings of the International Symposium on Archaeology of Medieval and Modern Ships of Iberian-Atlantic Tradition*, (Ed.), Francisco Alves, Instituto Portugués de Arqueologia, Lisbon: 381-403.

Castro, F. V., 2005. *The Pepper Wreck: A Portuguese Indiaman at the Mouth of the Targus River*. Texas A&M University, College Station.

Chakravarti, R., 1991. Horse Trade and Piracy at Tana (Thana, Maharashtra, India): Gleanings from Marco Polo. *Journal of the Economic and Social History of the Orient*, Vol. 34 (3): 159-182.

Chauhan, R. R. S., 1984. The Horse trade in Portuguese India. *Purabhilekh Puratatva*, Vol. 2 (1): 14-24.

Chauhan, R. R. S., 1988. Some interesting finding of sunken and wrecked ships in Indian Ocean during ancient and medieval periods. In *Marine Archaeology of Indian Ocean countries*, (Ed.), S. R. Rao, NIO, Goa: 14-16.

Child, A. M., 1995. Towards an understanding of the decomposition of bone in the archaeological environment. In *Biodeterioration and Biodegradation* 9, (Eds.), A. Bousher, M. Chandra and R. Edyvean), The 9th International Biodeterioration and Biodegradation Symposium, Institution of Chemical Engineers, Rugby, UK: 66-71.

Chirikure, S., Sinamai, A., Goagoses, E., Mubusisi, M., and Ndoro, W., 2010. Maritime Archaeology and Trans-Oceanic Trade: A Case Study of the Oranjemund Shipwreck Cargo, Namibia. *JMA*, Vol. 5 (1): 37-55.

Chittick, H. N., 1980. Stone anchor shanks in the Western Indian Ocean. *IJNA*, Vol. 9 (1): 73-76.

Cortesão, A., (Ed. & Tr.), 1990 and 2005. *The 'Suma Oriental' of Tomé Pires: An Account of the East, from the Red Sea to China and Japan.* 2 Vols, (Published in 1944, 1967, 1990), Asian Educational Services, New Delhi.

Costa, J. C., 1997. Archaeology: Treasure Trove. *Goa Today* (May 1997), Panaji: 24-30.

Costa, J. C., 2002. *The Heritage of Govapuri: A study of the Artefacts in and around the Pilar Seminary Museum.* Pilar Publications, Goa.

Cotsell, G., 1856. *Treatise on Ships' Anchors.* John Weale, London.

Cunha Rivara, J. H., (Ed.), *Archivo Portuguese Oriental, Nova Goa 1857-1877.* 6 Vols.

Curryer, B. N., 1999. *Anchors: An Illustrated History.* Chatham Publishing, London.

Danvers, F. C., 1988. *Portuguese in India,* Vol. 1 & 2, (Reprint). Asian Educational Services, New Delhi.

Darroch, C., 1986. The visionary shadow: a description and analysis of the armaments aboard the Santo Antõnio de Tanna. *Unpublished Master of Arts Thesis,* Texas A&M University, College Station, TX.

David, M. D., 1987. Historic Bassein. *Indica,* Vol. 24 (2): 89-104.

Delgado, J. P., (Ed.), 1997. *Encyclopaedia of Underwater and Maritime Archaeology.* British Museum Press, London.

De Kloguen, D. L. Cottineau., 1995. *A Historical sketch of Goa* (reprint). Asian Educational Services, Delhi.

Deloche, J., 1996. Iconographic Evidence on the Development of boat and ship structures in India (2nd century BC - 15th cent AD): a new approach. In *Tradition and Archaeology,* (Eds.), H. P. Ray and J. F. Salles, Manohar Publishers, New Delhi: 199-224.

Deloche, J., 2009. Stable-Ship of Tiruppudaimarudur and Tirukkurunkudi (South India). *IJHS,* Vol. 44 (4): 551-567.

Desai, P. B., 1965. Goa and Karnataka. *Jr. of the Karnataka University* (Social Sciences), Vol. 1: 12-27.

Desai, P. B., 1970. *A History of Karnataka.* Dharwad.

Desai, A. G., and Bodas, M. S., 1984. Occurrence of nepheline syenite around Murud-Janjira, Raigarh district, Maharashtra, India. *Current Science,* Vol. 53 (15): 775-777.

Desroches, J. P., 1998. Oriental ceramics and porcelains. In *Nossa Senhora dos Mãrtires The last Voyage,* (Ed.), A. A. Martins, Pavilhao de Portugal Expo'1998, Verbo: 229 - 251.

D'Silva, R. D., 1997. Ship-building in Portuguese Bassein 1534 -1739. In *Ship-building and Navigation in the Indian Ocean Region AD 1400-1800,* (Ed.), K. S. Mathew, Munshiram Manoharlal Publishers, New Delhi: 94-97.

De Souza, T. R., 1979. *Medieval Goa: A Socio-Economic History.* Concept Publication Company, New Delhi.

Disney, A., 1984. A 17th century Portuguese Proconsul: D. Miguel De Noronha, Count of Linhares. *Purabhilekh Puratatva*, Vol. 2 (2): 26-38.

Disney, A., 1990. The World of Long-Distance Voyaging in the Seventeenth Century: The Lisbon-Goa fleet of 1629 as a Case Study. In *Studies in Maritime History* (Ed.), K. S. Mathew, Pondicherry University Press, Pondicherry: 143-158.

Esteves, S., 1966. *Goa and its future.* Manaktalas, Bombay.

Evans, V., and Nutely, D., 1992. Hooked on anchors. *BAIMA*, Vol. 15 (2): 41-44.

Fenwick, V., 2000. Editorial, *IJNA*, Vol. 29 (1): 1-2.

Frost, H., 1985. Stone anchors; Criteria for a Corpus. *Thracia Pontica*, Vol. 3: 352-369.

Frost, H., 1989. 'Pyramidal' stone anchors: An inquiry. *Tropics*, Vol. 1: 97-114.

Fonseca, J. N., 1986. *A Historical and Archaeological Sketch of the City of Goa.* Asian Educational Services, New Delhi.

Furedi, H., and Walton, A. G., 1968. Transmission and Attenuated Total Reflection (ATR) Infrared Spectra of Bone and Collagen. *Appli. Spectros.,* Vol. 22 (1): 23-26.

Garlick, J. D., 1969. Buried bone: The experimental approach in the study of nitrogen content and blood group activity. In *Science in Archaeology* (2nd Edn), (Eds.), D. R. Brothwell and E. S. Higgs, Thames and Hudson, London: 503-512.

Gaur, A. S., 1992. Excavation at an Ancient Port in Gopakapattana: Goa. *JMA*, Vol. 3: 57-60.

Gaur, A. S., Sundaresh., Saxena, M. K, Sila Tripati., and Gudigar, P., 1997. Preliminary observations on an 18th century wreck at Poompuhar (east coast of India). *IJNA*, Vol. 26 (2): 118-126.

Gaur, A. S., Vora, K. H., Sundaresh., Sila Tripati., Gudigar, P., and Bandodker, S. N., 1998. Exploration of steam engine wrecks off Minicoy Island, Lakshadweep, India. *IJNA*, Vol. 27 (3): 225-236.

Gaur, A. S., Sundaresh., Sila Tripati., Gudigar, P., Vora, K. H., and Bandodker, S. N., 2001. A Group of 20 Stone anchors from the waters of Dwarka on the Gujarat Coast, India. *IJNA*, Vol. 30 (1): 95- 108.

Gaur, A. S., Sundaresh, Sila Tripati, and Bandodkar, S. N., 2002. Ring stone anchors from Gujarat, west coast of India. *Mariner's Mirror*, Vol. 88 (4): 390–404.

Gaur, A. S., and Sila Tripati., 2006. Marine archaeological explorations on the southwestern coast of Saurashtra, India. *JIOA*, Vol. 3: 81-89.

Gaur, A. S., Sundaresh., Sila Tripati., and Vora, K. H., 2007a. New evidence on the maritime activity at Dabhol on the Maharashtra coast. *Puratattva*, Vol. 37: 186-192.

Gaur, A. S., Sundaresh., and Sila Tripati., 2007b. Remains of the ancient ports and anchorage points at Miyani and Visawada, on the west coast of India: A study based on underwater investigations. *Mariner's Mirror*, Vol. 93 (4): 428-440.

Gaur, A. S., and Kerkar, R., 2007. Stone sculptures of goddesses on the boats from Goa, west coast of India. *BAIMA*, Vol. 31: 18-25.

Gaur, A. S., and Bhatt, B.K., 2008. Marine archaeological exploration on the western coast, Gulf of Khambhat. *Man and Environment*, Vol. 33 (2): 99-104.

Gaur, A. S., Sundaresh, Sila Tripati and Vora, K. H., 2009. Radiocarbon dates of the medieval period stone anchors from Dabhol, west coast of India. *Current Science*, Vol. 96 (2): 299-302

Gibb, H. A. R., 1929. *Ibn Batuta's Trevels in Asia and Africa (1325-1345 AD)*. George Routledge and Sons, London.

Godfrey, I. M., Ghisalberti, E. L., Beng, E. W., Byrne, L. T. and Richardson, G. W., 2002. The analysis of ivory from a marine environment. *Studies in Conservation*, Vol. 47: 29 45.

Gokul, A. R., Srinivasan, M. D., Gopalakrishnan, K., and Viswanathan, L. S., 1985. Geology, Stratigraphy, Structure and Tectonics. In *Earth Resources for Goa's Development*, (A Collection of papers submitted for the seminar held at Panaji, Goa on 18th-21st September 1981), Hyderabad.

Gomes, O. J. F., 2004. *Goa*. National Book Trust, New Delhi.

Green, J. N., 1973. The wreck of the Dutch East Indiaman the *Vergulde Draeck*, 1656. *IJNA*, Vol. 2 (2): 267-289.

Green, J. N., 1986. The Survey of the VOC *fluit Risdam* (1727), Malaysia. *IJNA*, Vol. 15 (2): 93-104.

Gudigar, P., and Sundaresh., 1992. Some important shipwrecks on Goa coast. *JMA*, Vol. 3: 48-53.

Gudigar, P., Sila Tripati., Gaur, A. S., and Sundaresh., 1995-96. Some important shipwrecks on Konkan coast. *JMA*, Vol. 5-6: 45-48.

Gudigar, P., 2000. *The Study of Ancient port of Gopakapattana, Field Session 1999-2000*, Technical Report No: NIO/TR-12/2000: 1-12.

Guerout, M., 2002. France. In *International Handbook of Underwater Archaeology*, (Eds.), C. V. Ruppe and J. F. Barstad, Kluwer Academic/Plenum Publishers, New York: 433-448.

Gune, V. T. (Ed.), 1979. *Gazetteer of the Union Territory of Goa, Daman and Diu, District Gazetteer*, Part-1, Goa.

Gune, V. T., 1990. Goa's Coastal and Overseas trade. In *Goa through the Ages - II* (Ed.), T. R. de Souza, Concept Publishing Company, New Delhi: 117-135.

Gune, V. T., 2002. Trade on the west coast of India before the conquest of Goa by the Portuguese. In *Studies in History of the Deccan (Medieval and Modern)*, Prof. A. R. Kulkurni Felicitation Volume, (Eds.), M. A. Nayeem, Aniruddha Ray and K. S. Mathew, Pragati Publications, New Delhi: 141-145.

Guy, J., 2004. Asian Trade and Exchange before 1600. In *Encounters: Asian and European exchange 1600 to 1800,* (Eds.), A. Jackson and A. Jaffer, V&A Publications, London: 52-69 and 88-89.

Heras, H., 1932. Pre-Portuguese remains in Portuguese India. *JBHS,* Vol. 4 (2): 1 - 56.

Hofmann, H., 1913. *The Basel Mission Industries.* Basel Mission Press, Mangalore.

Hourani, G. F., 1995. *Arab Seafaring in the Indian Ocean in Ancient and Early Medieval Times.* Princeton University Press, New Jersey.

Hunter, W. W., 1899. *A History of British India.* Longmans, Green & Co., London.

Indian Archaeology 2001-02 A Review. 2008, Archaeological Survey of India, New Delhi: 30-34.

Ingelman-Sundberg, C., 1976. The V.O.C. Ship '*Zeewijk*' 1727 Report on the 1976 survey of the site. *Australian Archaeology,* Vol. 5: 18-33.

Ingelman-Sundberg, C., 1977. The VOC ship *Zeewijk* lost off the Western Australian coast in 1727: An interim report on the first survey. *IJNA,* Vol. 6 (3): 225–231.

International Association of Wood Anatomists, IAWA list of microscopic features for hardwood identification. *IAWA Bulletin n.s.* 1989, Vol. 10: 221-332.

Iyer, S. D., Vora, K. H., and Gaur, A. S., 2006. Geological significance of stone anchors from Dwarka waters, Gujarat, India. In *Glimpses of Marine Archaeology in India. Proceedings of the Seventh Conference on Marine Archaeology of Indian Ocean Countries,* 6-7 October 2005, (Eds.), A. S. Gaur, and K. H. Vora, Society for Marine Archaeology, NIO, Goa: 78-81.

Iyer, S. D., Banerjee, G., Wagle, B. G., 1989. Statistical analysis of lineaments of Goa, India. *Current Science,* Vol. 58 (23): 1316-1319.

Iyer, S. D., and Wagle, B. G., 1987. Morphometric Analyses of the River Basins in Goa. *Geographical Review of India,* Vol. 49 (2):11-18.

Jane, F. W., 1970. *The Structure of Wood,* Adam and Charles Black, London.

Johnstone, P. F., and Robinson, D. S., 1993. The wreck of the 1848 Propeller Indiana: interim report. *IJNA,* Vol. 22 (3): 219-235.

Jones, J. W., 1863. *The Travels of Ludovico di Varthema in Egypt, Syria, Arabia Deserta and Arabia Felix in Persia, India and Ethiopia 1503-1508.* The Hakluyt Society, London.

Kamat, P., 2008. *Tarini and Tarvir-The Unique Boat Deities of Goa.* Goa.

Kamat, P., 1990. Historical Geography and Natural Resources. In *Goa through the Ages - II*, (Ed.), T. R. de Souza, Concept Publishing Company, New Delhi: 1-54.

Kamat, V., 2000, Trade and Commerce under Yusuf Adil Shah (1489-1510 AD). In *Goa: Trade and Commerce through the Ages*, (Ed.), S. K. Mhamai, Directorate of Archives and Archaeology, Panaji: 43-55.

Kapitan, G., 1994. Stone shank anchors of the Arab-Indian trade period – were they mooring anchors? *BAIMA*, Vol. 18 (2): 1-6.

Kennet, D., and Rao, J. V. P., 2001. The Early Historic Brick Temple at Chandor (Ancient Chandrapura), Goa. *SAS*, Vol. 17: 97-107.

Kimpton, G., 1992. Construction of replica anchor stocks and a corronade carriage for display of artefacts from HMS Sirius (1790). *BAIMA*, Vol. 16 (2): 31-38.

Krishnamacharlu, C. R., 1942. Siroda plates of Devaraja. *EI*, Vol. 24 (1937-38): 143-145.

Lambert, A., 2002. *Trincomalee - The last of Nelson's frigates.* Chatham Publishing, London.

Laval, de Pyrard., 1601-11. *Viagem de Francinsco Pyrard de Laval.* (Tr. J. H. de Rivara Cunha). Porto.

Lee, S., 1829. *Travels of Ibn Batuta*, (Reprint 2009). Cosimo Classics, New York.

Linschoten, J. H., 1903. *Voyage to Goa and Back, 1583-1592, with his account of the East Indies.* Archibald Constable and Co., Westminster, UK.

Lobato, M., 1997. Maritime trade from India to Mocambique - A study of Indo Portuguese Enterprise (16th to 17th centuries). In *Shipbuilding and Navigation in the Indian Ocean Region AD 1400-1800*, (Ed.), K. S. Mathew, Munshiram Manoharlal Publishers, New Delhi: 113-131.

Mascarenhas, A., 1987. *Goa from Prehistoric Times.* Holi Spirit Press, Vasco.

Martin, C., 2005. The *Adelaar*: a Dutch East-Indiaman Wrecked in 1728 off Barra, Outer Hebrides, Scotland. *IJNA*, Vol. 34 (2): 179-210.

Martins, Xavier, M., 1994. *Portuguese Shipping and Shipbuilding in Goa 1510-1780.* Unpublished Ph.D thesis, Goa University, Goa.

Martins, Xavier., 2008. Portuguese Shipbuilding in Bassein from 1534 to 1739: A Note. *Journal of the Vasco Da Gama Research Institute of Cochin*, Vol.1 (1): 89-99.

Mathew, K. M., 1988. *History of the Portuguese Navigation in India.* Mittal Publications, New Delhi.

Mathew, K. S., 1982. A Survey of the Portuguese Forts on west coast of India. *The Mythic Society Journal*, Vol. 73 (2): 55-72.

Mathew, K. S., 1990. Trade and Commerce in Sixteenth Century Goa. In *Goa through the Ages - II*, (Ed.), T. R. de Souza, Concept Publishing Company, New Delhi: 137-145.

Mathew, K. S., (Ed.), 1997. *Shipbuilding and Navigation in the Indian Ocean Region AD 1400-1800.* Munshiram Manoharlal Publishers, New Delhi.

Mathew, K. S., 2002. Life on board a Portuguese ship during the Sixteenth century. In *Studies in History of the Deccan (Medieval and Modern)*, Prof. A. R. Kulkurni Felicitation Volume (Eds.), M. A. Nayeem, Aniruddha Ray and K. S. Mathew, Pragati Publications, New Delhi: 162- 178.

May, S. R., 1986. The Valetta Archaeological Findings. *BAIMA*, Vol. 12: 27-35.

McCarthy, M., (Ed.), 1988. *Iron Ships and Steam Shipwrecks, (Papers from the First Australian Seminar on the Management of Iron Vessels and Steam Shipwrecks).* Western Australian Museum, Perth.

McConnell, D., 1965. Crystal chemistry of hydroxyapatite: Its relation to bone mineral. *AOB*, Vol. 10 (3): 421-431.

McCrindle, J. W., 1885. *Ancient India as Described by Ptolemy*, (Reprinted in 1985). Today and Tomorrow's Printers and Publishers, Delhi.

Merh, S. S., 1995. *Geology of Gujarat.* Geological Society of India, Bangalore.

Mookerji, R. K., 1912. *Indian Shipping: A History of the Seaborne Trade and Maritime Activity of the Indians from the Earliest Times.* Longmans, Green & Co, Bombay.

Morris, R., 1984. The 15th century hand-gun ex Association. *IJNA*, 13 (1): 76-77.

Moore, E., 1970. A suggested classification of Stone Wares of Martabani type. *Sarawak Museum Journal*, Vol. 18 (36-37): 1-78.

Moraes, G. M., 1990. *The Kadamba Kula, A History of Ancient and Mediaeval Karnataka.* Asian Educational Services, New Delhi.

Mustapa, S. A. H. B. Syed., 2005. Showcasing Maritime Heritage Artefacts for the Benefit of the Tourist Industry in Malaysia. *IJNA*, Vol. 34 (2): 211-215.

Nainar, S. M. H., 1942. *Arab Geographers knowledge of Southern India.* Islamic Series, No. 6, Madras University, Madras.

Nambirajan, M., and Gaur, A. S., 1997. River Ports and other Archaeological sites on the river banks of Goa. In *An integrated approach to Marine Archaeology*, (Ed.), S. R. Rao, NIO, Goa: 115-119.

Nash, M., 2002a. A survey of the steamship Tasman (1873-1883). *BAIMA*, Vol. 26: 83-90.

Nash, M., 2002b. The Sydney Cove shipwreck project. *IJNA*, Vol. 31 (1): 39-59.

Negi, B.S. and Purkayastha, S.K., 1985. Family Lauraceae. In *Indian Woods, their identification, properties and uses*, (Ed.), S. K. Purkayastha, The Controller of Publications, New Delhi.

Nicholson, R. A., 1996. Bone Degradation, Burial Medium and Species Representation: Debunking the Myths, an Experiment-based Approach. *JAS*, Vol. 23 (4): 513-533.

O'Connor, T. P., 1987. On the structure, chemistry and decay of bone, antler and ivory. In *Archaeological Bone, Antler and Ivory*, Occasional Papers No 5, (Eds.), K. Starling and D. Watkinson, The United Kingdom Institute for Conservation, London: 6-8.

Orser, C. E., (Ed.) 2002. *Encyclopedia of Historical Archaeology*. Routledge, London.

Osbourne, A., 1941. *Modern Marine Engineer's Manual*. Cornell Maritime Press, New York.

Panikkar, K. M., 1997. *Malabar and the Portuguese*. Voice of India, New Delhi.

Parthesius, R., Millar, K., Devendra, S., and Green, J., (Eds.) 2003. *Sri Lanka Maritime Archaeological Unit Report on the Avondster Project 2001-2002*. Amsterdams Historisch Museum, The Netherlands.

Pate, F. D., Hutton, J. T. and Norrish, K., 1989. Ionic exchange between soil solution and bone: towards a predictive model. *Appli. Geochemis.*, Vol. 4 (3): 303-316.

Pathak, M. C., Hashimi, N. H., and Nigam, R., 1991. Relevance of topography and sediment in shipwreck exploration - Examples from Goa waters. In *Recent Advances in Marine Archaeology*, (Ed.), S. R. Rao, National Institute of Oceanography, Goa: 135-139.

Penniman, T. K., 1952. Pictures of Ivory and other animal teeth bone and antler. In *Occasional Paper on Technology No 5*, (Eds.), T. K. Penniman and B. M. Blackwood, Pitt Rivers Museum University of Oxford, Oxford: 13-20.

Piercy, R. C. M., 1977. Mombasa wreck excavation: Preliminary report 1977. *IJNA*, Vol. 6 (4): 331- 347.

Piercy, R. C. M., 1978. Mombasa wreck excavation, Second Preliminary report, 1978. *IJNA*, Vol. 7(4): 301-319.

Piercy, R. C. M., 1981. Mombasa wreck excavation, Fourth preliminary report, 1980. *IJNA*, Vol. 10 (2): 109-118.

Pinto, C., 1994. *Trade and Finance in Portuguese India*. Concept Publishing Company, New Delhi.

Pires, Tom., 1944. *The Suma Orientale*, (Ed.), A. Cortesão, Vols. I and II, Hakluyt Society, London.

Pires, Tom., 1967. *The Suma Oriental*, Edited and Translated by Armando Cortesão, Vol. I, Nendeln/Liechtenstein.

Pissurlencar, P. S., 1938. *Inscricoes Pre-Portuguesa da Goa, Orienta Portuguesa.* Goa Archives, Panaji.

Pissurlencar, P. S., 1857-1877. *Regimento das Fortalezas da India.* Vol. II, Goa.

Puech, A., 1984. *The use of anchors in offshore petroleum operations.* Gulf Publishing Company, Houston.

Purkayastha, S. K., and Lal, K, 1972. Family Combretacea, In *Indian Woods, Their identification, properties and Uses,* (Ed.), K. R. Rao and S. K. Purkayastha, Manager of Publications, New Delhi.

Purkayastha, S. K., 1982. *Indian Woods their identification, properties and uses,* The controller of publications, Delhi.

Purkayastha, S. K., 1985. *Indian Woods their identification, properties and uses,* The controller of publications, Delhi.

Qaisar, A. J., 1982. *The Indian response to European Technology and Culture (AD 1498-1707).* Oxford University Press, New Delhi.

Qualls, C., 1985. Early Mesopotamian boats and the Gulf. Unpublished paper presented at the Bahrain Historical and Archaeological Society.

Raban, A., 1990. Medieval anchors from the Red Sea. *IJNA,* Vol. 19 (4): 299-306.

Raban, A., 2000. Three-hole composite stone anchors from a medieval context at Caesarea Maritima, Israel. *IJNA,* Vol. 29 (2): 260-272.

Radhakrishna, B.P., Vaidyanadhan, R., 1994. *Geology of Karnataka.* Geological Society of India, Bangalore.

Rajagopalan, S., 1987. *Old Goa.* Archaeological Survey of India, New Delhi.

Raghaviah, J., 1990. *Basel Mission Industries in Malabar and South Canara (1834-1914).* Gian Publishing House, New Delhi.

Rao, R. K., and Juneja, K. B. S., 1992. *Field identification of fifty important timbers of India.* Indian Council of Forestry Research and Education, New Forest, Dehra Dun.

Rao, N. L., 1941-42. A Note on the Siroda Plates of (Bhoja) Devaraja. *EI,* Vol. 26: 337-340, (Reprint 1985).

Rao, S. R., 1970. Shipping in Ancient India. In *Indian's Contribution to World Thought and Culture, Vivekananda Commemoration Volume,* (Ed.), L. Chandra, Vivekananda Rock Memorial Committee, Madras: 83-107.

Rao, S. R., Vora, K. H., Gaur, A. S., Gudigar, P., Sundaresh., and Sila Tripati., 1995-96. A Preliminary Shipwreck Exploration off Lakshadweep. *JMA,* Vol. 5-6: 23-32.

Rogers, E. M., 1992. Boat reliefs in the tomb of Ti and Mastaba of Mereruka. *The INA Quarterly*, Vol. 19 (3): 8-11.

Roy, A. C., 1972. *A History of the Mughal Navy and Naval warfare.* The World Press Ltd, Calcutta.

Rubin, N., 1971. The Anchor. *NRJ,* Vol. 18: 230-250.

SanilKumar, V., Pathak, K. C., Pednekar, P., Raju, N. S. N., and R. Gowthaman., 2006. Coastal processes along the Indian coastline. *Current Science*, Vol. 91(4): 530-536.

Saraswati, A. R., 1925. Political Maxims of Emperor Poet: Krishnaraya. *JIH*, Vol. 6 (3): 72.

Sassoon, H., 1981. Ceramics from the wreck of a Portuguese ship at Mombasa. *Azania*, Vol. 16: 97-130.

Sassoon, H., 1991. The Portuguese Faience. *INA Newsletter*, Vol. 18 (2): 22-24.

Scammell, G. V., 1982. European Seamanship in the Great Age of Discovery. *Mariner's Mirror*, Vol. 68 (4): 10.

Scammell, G. V., 1995. *Ships, Oceans, Empire: Studies in European Maritime and Colonial History 1400-1750* . Hamphire.

Schoff, W. H., 1974. *The Periplus of the Erythraean Sea*, (2nd edition). Oriental Books, New Delhi.

Sewel, R., 1900. *A Forgotten Empire.* London.

Sharma, S. N., Kumar, S., Tiwari, M. C. and Sekhar, A. C., 1974. Timber for Boat and Shipbuilding. *Jr. Timb. Dev. Assoc. India*, Vol. 20 (1): 9-18.

Shastry, B. S., 1973. The System of *Cartazes. JIH,* Vol. 51: 297-300.

Shastry, B. S., 1981. *Studies in Indo-Portuguese History.* IBH Prakashana, Bangalore.

Shirodkar, P. P., 1988. India and Mozambique: Centuries old interaction. *Purabhilekh-Purtatva*, Goa Historical Archives, Vol. 6 (1): 35-62.

Shirodkar, P. P., 1990, Survey of the ancient Silahara-Kadamba port of Gopakapattana. *JMA*, Vol. 1: 52-54.

Sila Tripati., and Gaur, A. S., 1997. Stone anchors from Sindhudurg Fort on the west coast of India. *IJNA*, Vol. 26 (1): 51-57.

Sila Tripati., Gaur, A. S., Sundaresh., and Bandodker, S. N., 1998a. Historical period Stone anchors from Vijaydurg on the west coast of India. *BAIMA,* Vol. 22: 1-8.

Sila Tripati., Saxena, M. K., Sundaresh., Gudigar, P., and Bandodkar, S. N., 1998b. Marine Archaeological Exploration and Excavation of Vijaydurg - a naval base of the Maratha Period, Maharashtra, on the west coast of India. *IJNA*, Vol. 27 (1): 51-63.

Sila Tripati., 1999. *Exploration and Excavation of shipwrecks in Goa waters, Progress and Final Report,* Technical Report No: NIO/TR-1/1999: 1-35.

Sila Tripati., Gaur, A. S., Sundaresh., and Bandodker, S. N., 2001. Exploration for Shipwrecks off Sunchi Reef, Goa, West Coast of India. *World Archaeology,* (Shipwrecks) Vol. 32 (3): 355-367.

Sila Tripati., and Gudigar, P., 2001. Shipwreck archaeology of the Lakshadweep Islands, West coast of India. *IJNA,* Vol. 30 (1): 37-47.

Sila Tripati., 2002a. *Exploration and Excavation of shipwrecks in Goa and adjoining waters 2001-2002,* Technical Report No: NIO/TR-8/2002: 1-27.

Sila Tripati., 2002b. Pyramidal anchor stone from Baga waters of Goa, west coast of India. *The Mariner's Mirror,* Vol. 88 (1): 6-13.

Sila Tripati., Gaur, A. S., Sundaresh., and Bandodker, S. N., 2002. Marine Archaeological explorations off Goa, India. *IJMH,* Vol. 14 (2): 225-245.

Sila Tripati., Gaur, A. S., and Sundaresh., 2003a. Anchors from Goa waters, Central West Coast of India: Remains of Goa's Overseas Trade Contacts with Arabian countries and Portugal. *BAIMA,* Vol. 27: 97-106.

Sila Tripati., Sundaresh., Gaur, A. S., Gudigar, P., and Bandodker, S. N., 2003b. Exploration of Basel Mission Company shipwreck remains at St George's Reef off Goa, West Coast of India: impact of the Basel Mission Co. on society and culture. *IJNA,* Vol. 32 (1): 111-120.

Sila Tripati., Parthiban, G., Vora, K. H., Sundaresh., and Bandodker, S. N., 2003c. Lead ingots from a shipwreck off Poompuhar, Tamil Nadu, East Coast of India: evidence for overseas trade and their significance. *IJNA,* Vol. 32 (2): 225-237.

Sila Tripati., 2004. Exploration of Portuguese shipwreck in Goa waters of India: With special reference to Sunchi Reef. In *The Application of Recent Advances in Underwater Detection and Survey Techniques to Underwater Archaeology,* (Eds), Tuncay Akal, Robert D. Ballard and George F. Bass, Ulu Burun Publishing, INA, Bodrum: 19-29.

Sila Tripati., Gaur, A. S., Sundaresh., and Vora, K. H., 2004. Shipwreck archaeology of Goa: Evidence of maritime contacts with other countries. *Current Science,* Vol. 86 (9): 1238-1245.

Sila Tripati., M. Sujatha, Rao, R. V., and Rao, K. S., 2005a. Use of timber in shipbuilding industry: Identification and analysis of timber from shipwrecks off Goa coast, India. *Current Science,* Vol. 89 (6): 1022-1027.

Sila Tripati., Manikfan, A., and Mohamed, M., 2005b. An Indo-Arabian type of stone anchor from Kannur, Kerala, west coast of India. *IJNA,* Vol. 34 (1): 131-137.

Sila Tripati., Gaur, A. S., Sundaresh., 2006a. Exploration of a Portuguese shipwreck in Goa waters, western coast of India. *BAIMA*, Vol. 30: 127-136.

Sila Tripati., Gaur, A. S., Sundaresh., and Vora, K. H., 2006b. Maritime archaeological Explorations of Goa: Findings and Interpretations. In *Glimpses of Marine Archaeology in India*, (Eds.), A. S. Gaur and K. H. Vora, Society for Marine Archaeology, NIO, Goa: 35-41.

Sila Tripati., 2006a. Ships on Hero Stones from the West Coast of India. *IJNA*, Vol. 35 (1): 88-96.

Sila Tripati., 2006b. *Exploration and Excavation of shipwrecks in Goa and adjoining waters 2005-2006*, Technical Report No: NIO/TR-7/2006: 1-28.

Sila Tripati., and Godfrey, Ian., 2007. Studies on the elephant tusks and hippopotamus teeth collected from the early 17th century Portuguese shipwreck off Goa, West coast of India: Evidence of maritime trade between Goa, Portugal and African countries. *Current Science*, Vol. 92 (3): 332-339.

Sila Tripati., and Patnaik, A. P., 2008. Stone anchors along the coast of Chilika Lake: New light on the maritime activities of Orissa, India. *Current Science*, Vol. 94 (3): 386-390.

Sila Tripati., Sundaresh., Gaur, A. S., and Vora, K. H., 2008. Sunken ships and a Long Lost Port. In *Goa Aparanta – Land beyond the End*, (Ed.), Victor Rangel Ribeiro, Goa Publications Pvt Ltd, Goa: 46-47.

Sila Tripati, 2008. *Exploration and Excavation of shipwrecks in Goa and adjoining waters 2007-2008*, Technical Report No: NIO/TR-1/2008: 1-19.

Sila Tripati., Rao, B. R., Shashikala, S., Rao, R. V, and Khedekar, V., 2009. Analysis of timber and coating material on an iron anchor recovered off Aguada Bay, Goa. *Current Science*, Vol. 97 (9): 1361- 1368.

Sila Tripati., 2009. Stone anchors from Minicoy Island, Lakshadweep, India. *IJNA*, Vol. 38 (2): 406-412.

Sila Tripati., Gaur, A. S., and Sundaresh., 2010a. Exploration of a steamship wreck off Amee shoals, Goa, India: A preliminary report. *IJNA*, Vol. 39 (1): 182-189.

Sila Tripati, Mudholkar, A., Vora, K. H., Rao, B. R., Gaur, A. S., Sundaresh., 2010b. Geochemical and mineralogical analysis of stone anchors from west coast of India: provenance study using thin sections, XRF and SEM-EDS. *Journal of Archaeological Science*, Vol. 37 (8): 1999-2009.

Sillen, A., 1989. Diagenesis of the inorganic phase of cortical bone. In *The chemistry of prehistoric human bone*, (Ed.), T. D. Price, School of American Research Seminar Series, Cambridge University Press, Cambridge: 211-229.

Souter, C., 1998. Stone anchors near Black Fort, Galle, Sri Lanka. *IJNA*, Vol. 27 (4): 331-342.

Stephen, S. J., 1997. Portuguese Nau: A study of the cargo ship in the Indian Ocean region during the Sixteenth century. In *Shipbuilding and Navigation in the Indian Ocean Region AD 1400-1800*, (Ed.), K. S. Mathew, Munshiram Manoharlal Publishers, New Delhi: 62-81.

Stern, W. L., 1954. Comparative anatomy of xylem and phylogeny of Lauraceae. *Trop. Woods*, Vol. 100: 1-73.

Sundaresh., Gaur, A. S., Gudigar, P., Sila Tripati., Vora, K. H., and Bandodker, S. N., 1999. Stone anchors from the Okhamandala Region, Gujarat Coast, India. *IJNA*, Vol. 28 (3): 229-252.

Sundaresh., Gaur, A. S., Sila Tripati., Gudigar, P., and Bandodker, S. N., 2002. Stone anchors from Bet Dwarka Island, Gujarat coast, India: Significance to historical period maritime activities. *BAIMA,* Vol. 26: 43-50.

Thakkar, M., 1991a. Portuguese shipwreck on the Goa coast. *JMA*, Vol. 2: 107-109.

Thakkar, M., 1991b. Search for shipwrecks in Goa water. In *Recent Advances in Marine Archaeology*, (Ed.), S. R. Rao, NIO, Goa: 195 -197.

Turner, M., 1988. *Shipwrecks and Salvage*: In South Africa - 1505 to the Present. C.Struik, Struik House, Cape Town.

Tzalas, H., 1999. Were the 'Pyramidal stone weights' of Zea used as anchors? *Troipcs,* Vol. 5: 429-447.

Upham, N. E., 1983. *Anchors.* Shire Publications, Bucks.

Vaidyanadhan, R., 1968. *Publication of the CSIR.* New Delhi.

Vander Pijl-Ketel, C., and Kist, J. B., (Eds.) 1982. *The Ceramic load of the Witte Leeuw (1613)*, Amsterdam: Rijksmuseum.

Veerayya, M., Murty, C. S., and Varadachari, V. V. R., 1981. Sediment distribution in the offshore region of Goa. *IJMS*, Vol.10 (7): 332-336.

Vliet van, G. J. C. M., 1979. Wood Anatomy of the Combretaceae. *BLUMEA*, Vol. 25: 141-223.

Vosmer, T., 1999. Indo-Arabian Stone Anchors in the western Indian Ocean and Arabian Sea. *AAE*, Vol. 10: 248-263.

Wagle, B. G., 1982. Geomorphology of the Goa Coast. *Proceedings of Indian Academy of Sciences (Earth Planetary Sciences)*, Vol. 91(2): 105 - 117.

Wagle, B. G., Gujar, A. R., Subrahmanyam, V., and Mislankar, P. G., 1988. Seabed Surveys of Mormugoa Harbour, Central West Coast of India. *IJMS*, Vol. 17 (1): 59 - 62.

Warmington, E. H., 1974. *The Commerce between the Roman Empire and India.* Vikas Publishing House, New Delhi.

Webster, R., 1958. Ivory, bone and horn. *The Gemmologist,* Vol. 27 (1): 91-98.

Weech, W. N., 1961. *History of the World,* Odhams Press, London.

Whitehouse, D., 1970. Excavations at Siraf. Third Interim Report. *Iran,* Vol. 8: 1-17.

Willoughby, F. P., 1991. The Martaban Jars. *INA News Letter,* Vol.18 (2): 25-27

Winkler, H. A., 1938. *Rock-drawings of Southern Egypt,* Vol. l, Egyptian Exploration Society, London.

Index

D

E

F

G

N

O

P

Q

R

S

❐❐❐